Laws and Order in Eighteenth-Century Chemistry

TABLE DES DIFFERENTS RAPPORTS
observés entre différentes substances.

Mem. de l'Acad. 1718 Pl. 8

Esprits acides.
Acide du sel marin.
Acide nitreux.
Acide vitriolique.
Sel alcali fixe.
Sel alcali volatil.
Terre absorbante.
SM Substances metalliques.
Mercure.
Regule d'Antimoine.
Or.
Argent.
Cuivre.
Fer.
Plomb.
Etain.
Zinc.
PC Pierre Calaminaire.
Soufre mineral.
Principe huileux ou Soufre Principe.
Esprit de vinaigre.
Eau.
Sel.
Esprit de vin et Esprits ardents.

Geoffroy's *Table des Differents Rapports* (1718).

List of substances mentioned.

Esprit acides: Acid spirits – i.e. acids in general

Acide du sel marin: Acid of marine salt (hydrochloric acid)

Acide nitreux: Nitrous acid (modern nitric acid)

Acide vitriolique: Vitriolic acid (sulphuric acid)

Sel alcali fixe: Fixed alkaline salt (sodium or potassium hydroxide)

Sel alcali volatil: Volatile alkaline salt (ammonia)

Terre absorbante: Absorbent earth (generally chalk, i.e. calcium carbonate)

Substances metalliques: Metallic substances

Mercure: Mercury

Regule d'Antimoine: Regulus of antimony (metallic antimony)

Or: Gold

Argent: Silver

Cuivre: Copper

Fer: Iron

Plomb: Lead

Etain: Tin

Zinc: Zinc

Pierre Calaminaire: Calamine (zinc carbonate)

Soufre mineral: Mineral sulphur

Principe huileux ou Soufre Principe: Oily principle or Sulphur principle – i.e. phlogiston

Esprit de vinaigre: Spirit of vinegar (acetic acid)

Eau: Water

Sel: Salt

Esprit de vin et Esprits ardents: Spirit of wine and ardent Spirits (ethyl alcohol with various congeners)

Laws and Order in Eighteenth-Century Chemistry

ALISTAIR DUNCAN

Formerly Reader in History and Philosophy of Science, Loughborough University of Technology

CLARENDON PRESS . OXFORD
1996

Oxford University Press, Walton Street, Oxford OX2 6DP
Oxford New York
Athens Auckland Bangkok Bombay
Calcutta Cape Town Dar es Salaam Delhi
Florence Hong Kong Istanbul Karachi
Kuala Lumpur Madras Madrid Melbourne
Mexico City Nairobi Paris Singapore
Taipei Tokyo Toronto
and associated companies in
Berlin Ibadan

Oxford is a trade mark of Oxford University Press

Published in the United States
by Oxford University Press Inc., New York

A catalogue record for this book is available from the British Library

Library of Congress Cataloging in Publication Data
(Data available)

ISBN 0 19 8558066

Typeset by Footnote Graphics, Warminster
Printed in Great Britain by
Biddles Ltd
Guildford and Kings Lynn

Preface

A book which is about the internal development of chemistry inevitably seems old-fashioned nowadays. For the past twenty years or so, the front line of the history of science has attacked the social history of scientists and of scientific institutions and the social functions of science in different periods. Hence we have a much broader and deeper understanding of the context in which science developed and of the people who developed it than we had a quarter of a century ago. Yet one of the purposes of historians of science must surely still be to interpret the scientific ideas themselves within that context. I therefore make no apology for writing in a way which may seem old-fashioned. It must also be emphasized that I am writing here about a single strand in eighteenth-century chemistry, and so that I have necessarily neglected many others which are essential parts of the whole. Also, although I have included an outline of seventeenth-century work on the nature and methods of science (as we now call it), it is intended merely as an outline to form a background to eighteenth-century science, and not by any means as a thorough discussion.

In any case, it is no longer possible to write exactly as we used to write: awareness that each chemist was part of society at a particular time and place, educated in a particular way, working within a particular social and institutional structure and in a particular country, within a particular universe of discourse and with particular ambitions and criteria of success, must always inform our understanding of his (it was almost always his) scientific achievements and theories.

Another immense advance in the history of science has been made by the Newton industry—that is, by the study of Newton's own thought and unpublished writing, especially on alchemy, rather than by seeing him from a nineteenth-century viewpoint. Indeed, Newton was a powerful influence on the concepts of eighteenth-century chemistry. No account of it can neglect him. Nevertheless, most chemists were on the whole less concerned with the subtleties of philosophy and more concerned with finding theories which would help them to understand what they saw happening in their laboratories. It is obvious that the notion of chemical attraction, and the originally quite different notion of chemical affinity which eventually merged into it, were very important to eighteenth-century chemists, who often mention them. However, because the needs and aims and inhibitions of chemists were different from those of physicists and abstract philosophers, the notion of chemical attraction is different in many ways from the notion of attraction in physics, though much influenced by it.

Most work on Newtonianism in the eighteenth century has concentrated on the physicists, and on their explanations of chemical phenomena. This book is an attempt to see things from the chemists' viewpoint; and I have therefore avoided undue emphasis on Newtonianism in order to give a clearer account of the distinctively chemical concepts, which were increasingly independent of the physicists' and mathematicians' concepts. Indeed, I have tried to bring out the ways in which chemistry achieved autonomy as a discipline in its own right.

The material on which this book is based is drawn largely from a thesis accepted by the University of London for the degree of Doctor of Philosophy in 1971; but in the interim I have begun to understand what it all means. It is pleasant to be able to acknowledge the help which I have received over the years from discussions with the late Professor D. McKie, with Dr W. A. Smeaton, with my former colleague Dr J. G. Smith, and with many other fellow historians. I am also grateful for the suggestions made by the anonymous referees who read the book in draft. However, its shortcomings are entirely mine.

I have quoted from contemporary translations where they are available, or available modern translations in the absence of contemporary translations; but unless otherwise indicated translations are my own.

It may be misleading to give twentieth-century equivalents for seventeenth- or eighteenth-century chemical names of substances. Even in the eighteenth century substances were often impure by modern standards. There are a number of instances of different substances which were thought to be the same in the eighteenth century, or of substances which were thought to be different but turned out to be the same, quite apart from changing notions of which substances were elementary and which were compound, from vague mixtures which were regarded as distinct substances, and indeed from substances such as phlogiston or caloric which are not now thought to be substantial at all. Nevertheless, to assist the reader I have added a modern name the first time an obsolete name occurs in a passage. However, it should be understood that the modern names are only rough equivalents to the older names.

Loughborough
May 1995 A.M.D.

Contents

1
The background of eighteenth-century chemistry

Introduction

In the study of eighteenth-century chemistry most attention has been paid to the development of understanding of the nature of various substances. Less attention has been paid to concepts of the nature of chemical change, and its causes and mechanisms. Yet if we are to understand fully the chemistry of the eighteenth century, and the set of concepts and attitudes which were inherited by chemists of the nineteenth century, then causes and mechanisms of chemical change are even more important than theories of the nature of the substances involved in it. The notion of chemical affinity or attraction was a central feature of eighteenth-century chemistry, as many chemists realized and stated; and the ways in which that notion was used provide the main theme of this book. In this introductory chapter, I shall sketch some of the attitudes and assumptions which form the context within which the notion of affinity or attraction developed, and some aspects of the social conditions of chemists at that time.

On the other hand, we must beware of attributing to chemists the beliefs and assumptions that were held by physicists. Natural philosophers of the eighteenth century speculated boldly and variously about the shapes, sizes, and behaviour of the particles of which they supposed matter to consist.[1] However, for reasons which will be suggested below, chemists often did not feel entitled to follow such speculations. If we are to understand their work we must put ourselves into their position and try to avoid assuming that such speculations were part of their thought. Also, there is another large area of modern knowledge which we must avoid attributing to early eighteenth-century chemists. Isaac Newton was one of the most powerful influences on eighteenth-century chemists, as on natural philosophers generally. In recent years a great deal has been learnt about Newton's own thinking, especially about alchemy and to some extent about theology. Some of that research will be mentioned in my next chapter. Nevertheless, little of his ideas about such topics was known to the eighteenth century, and we must be careful not to allow our knowledge of Newton's unpublished thought to affect our interpretation of eighteenth-century chemists who did not possess that knowledge. They saw Newton quite differently.

Until the eighteenth century, chemistry in Europe was not generally perceived as an academically respectable pursuit. While mathematics, astronomy, and physics had high prestige and were proper activities for scholars, chemistry was associated with dark and smoky laboratories, vulgar trades, and the furtive and occult practices of alchemy. It was also the concern of medical men, in two ways. First, it was involved in the preparation of medicines; and second, it had been introduced into chemical explanations of human physiology. However, association with medicine had not brought to chemistry acceptance as a respectable branch of philosophy. Chemists of course did not share this view of their activities. However, they generally showed respect and even in some cases a degree of humility towards the more prestigious disciplines, and felt obliged to accept their views on such subjects as the nature of matter, even if they were not really appropriate or helpful in chemistry. I do not mean to imply that chemists had some sort of inferiority complex about the more mathematical branches of what we should now call science: rather, chemistry was quite a different kind of activity. Seventeenth-century chemists were generally content to accept the status of their work, if they thought about it at all: their aims were not the same as those of astronomy or physics. However, in the eighteenth century many chemists felt the need to establish the public image of their art as being on an equal footing with such branches of natural philosophy as astronomy and mechanics.

In such textbooks of chemistry as there were, the authors were impelled at least to pay lip-service to philosophically respectable theories, such as Aristotle's theory of the four elements. During the late seventeenth and the early eighteenth centuries, however, chemistry was in the process of becoming an autonomous, prestigious, and successful branch of philosophy. Chemists were able to move away from the ideas imposed on them by physicists and to develop their own set of concepts which suited their own needs. Indeed, it was commonplace in the mid-eighteenth century for chemists to record with pride that not long before (apparently at the turn of the century) chemistry had emerged from the darkness of magic and superstition into a new, successful period of rational enlightenment.

For example, the *Nouveau cours de chymie* of 1723, usually attributed to Senac, in a long 'Discours historique', has the following remark:

For a very long time Chemistry was only empirical. It was noticed that certain mixtures produced certain effects; the reason was not sought for. Hence the first chemical books include only vague experiments. It is perhaps for that reason that Chemistry has not been given the esteem that it merited. It has been looked on as an art without art, in which no fixed point for guidance was ever seen. Eventually there was an attempt to bring to it the illumination of Physics, but with little success.[2]

The *Nouveau cours de chymie* is somewhat ahead of its time; but the same belief that chemistry had recently emerged from the darkness of irrationality into a proper state of enlightenment is frequently expressed in the third quarter of the century. In Margraf's *Opuscules chymiques*, translated into French in 1762, the point is emphasized in the preface.

Plunged formerly into the darkness of Scholasticism, disfigured by the emblems and hieroglyphs under which it was masked, chemistry served only to flatter the arrogant vanity of a very small number of men, or to feed by vain hopes the greed of some others. If there has emerged from the midst of this chaos with the help of lucky interpretations or endeavours some beam of light, far from drawing any brilliance from the darkness from which it came, this trembling light guided only a few privileged souls, still not bold enough to form from the combination of their experiments a body of doctrine.

But hardly had the more enterprising chemists got their system built, than, all shapeless as it inevitably was, it served as a centre on which the various experiments, previously scattered, relied. Chemistry was no longer confounded with magic, or empiricism; it was no longer the art of working wonders, while enriching the Adepts, or curing by a universal remedy all human infirmities; it became the sure torch of physics, its faithful companion and perhaps its mistress. From that time on it was almost equally shameful to conceal one's works or to pass for an alchemist.[3]

Evidently what Margraf saw as the great advance which had brought chemistry into the same class as physics was the sorting of a shapeless mass of experimental results into a theoretical system. Yet, as will be shown below, the nastiest accusation which could be made against a chemist of the mid-eighteenth century was that he had produced a system which was not based on experiment.

The point is made by Macquer in the 'Discours préliminaire' which introduces his influential chemical dictionary of 1766. After referring with respect to Stahl and Boerhaave, he writes:

If the theories of the great men of whom we have just spoken are capable of contributing infinitely to the advancement of chemistry, making us perceive the causes and the relationships of all the phenomena of that science, we must also acknowledge that they can produce a quite contrary effect, when one surrenders to them with too much confidence, and when one extends their use beyond their limits. Theory can be useful only as far as it is born from experiments already carried out, or as it shows us those which are to be carried out. For the reasoning is in a sense the organ of sight of the Physicist, but experiment is his touch, and this last sense should continually correct for him the errors to which the former is only too liable. If experiment which is in no way directed by theory is always a blind groping, theory without experiment is never more than a deceptive and undependable glimpse. Thus it is certain that the most important discoveries which have been made in Chemistry are due only to the combination of these two great resources.

A few lines further on he speaks of 'the veritable epoch at which we have seen the disappearance of the barbarous jargon of the Schools, the illusions of judicial Astrology, the extravagances of Alchemy, which were only chymerical speculations destitute of proof, or confused heaps of facts which proved nothing'.[4]

James Keir, in his English translation of Macquer's dictionary, put it more strongly:

Since Chemistry is delivered from the obscure jargon of alchemists, and from the narrow limits of one profession, namely, the medical, from which it has received much of its advancement; since it is now justly considered as one of the most important branches of natural philosophy; since it has become a more general object of study, and has lately been successfully cultivated at home, but especially abroad, by persons who employ the advantages attending rank, opulence, leisure, and philosophical minds, to the acquisition of knowledge, and improvement of useful arts; we may hope, that the number of chemical facts may be daily increased; that those discovered may be further ascertained and illustrated; and that some person of eminently perspicuous genius may unfold the laws and causes of chemical combination and analysis, collect and methodise the scattered facts, and form the several parts of Chemistry into one regular, connected science.[5]

Condorcet in his *Éloge* of Macquer, who died in 1784, summed up his position, though he seems to put the rationalization of chemistry rather later than Macquer himself had done. Writing as Secretary of the *Académie Royale*, Condorcet was a mathematician looking at chemistry from the outside.

M. Macquer found himself placed at an epoch at which chemistry began to deliver itself from the dreams of the alchemists, with which the works of the restorers of this science are still infected; but clarity, method, were an unknown virtue in the books which dealt with it: a residue of Cartesianism added to the obscurity of the science, while overloading it with pretended mechanical explanations.[6]

Another English example of this view is provided by the lectures, published in 1771, of Henry Pemberton, Professor of Physic (i.e. medicine) at Gresham College. After attacking the absurdity of such followers of Paracelsus as Van Helmont, he continues:

These men's absurdities, as well as the ridiculous original of the art, would naturally prejudice men of understanding against it; but as it has at length been also cultivated by men of more sober minds, it has surmounted these prejudices; and being found to have actually supplied us with many valuable remedies, a just distinction has been made between the art itself, and the follies of its professors; and such chemical preparations as are found to have real use in physic, are universally held in due esteem, without being either extravagantly extolled by one set of men, or unreasonably depreciated by another.[7]

He goes on to praise the chemical work of Boyle and Newton, to whom he evidently attributes the dawn of the new age in chemistry.

In a later generation, and perhaps with a foreshortened view, Thomson in his *History of chemistry* of 1830–1 has a chapter called 'Of the Attempts to Establish a Theory in Chemistry', in which he calls Becher the first to attempt to construct a theory of chemistry and gives favourable mention to a number of early- and mid-eighteenth-century chemists, including Macquer. In particular he identifies his fellow Scot William Cullen as having been the first to teach chemistry as a science independent of medicine. However, in an earlier chapter he had called Nicholas Lemery 'the first Frenchman who completely stripped chemistry of its mysticism, and presented it to the world in all its native simplicity'.[8]

To sum up, chemists of the mid-eighteenth-century—and later—presented an earlier stage of chemistry as having been mystical, obscure, and deluded, or as consisting merely of experiments without an organizing theory. They further presented the chemistry of their own era as having emerged at some time in the later seventeenth or early eighteenth century from that obscurity, and having become a rational, organized science, with a theoretical basis, on a par with such mathematically based branches of natural philosophy as astronomy or Newtonian mechanics. At the same time, they insisted that theory must be based on experiment and not imagined *a priori*.

The motives for this presentation of a new image of chemistry are fairly evident. Natural philosophy had become, as Golinski neatly expresses it, 'public knowledge'.[9] Natural philosophers were seen as proper subjects for royal or noble patronage, especially in France and Germany. As commerce and manufactures developed, the professions which supported them established a position of prestige in society, alongside the traditional professions of the church, the law, and the armed forces. Natural philosophers, though not yet constituting a profession, were attaining a similar status. Chemists were successfully asserting a claim to share that status. The situation is clearly stated by Cullen, who justified the inclusion of philosophy, that is theory, as well as practical chemistry in his lectures for medical students, so that chemistry might be recognized as 'a considerable part of Natural Philosophy capable of being applied to very important purposes of Society'.[10]

The difference between chemistry and physics is not only that the apparatus and the jars and bottles and the colours and smells are different: the chemist in trying to understand chemical change is generally trying to answer different questions from those of the physicist, and eventually has to answer them by using methods and concepts and ways of thought which are also different. Superficially, the subject-matter of chemistry is more obvious and more easily observed, indeed often more spectacular, than that of the other physical sciences. Chemical changes are going on round us all the time, in burning, rotting, rusting, and many other natural processes. These changes are surprising, if we stop to consider them; for they involve the substance that burns or rots or rusts changing into something completely different. Any explanation of such visible changes must depend on imagining mechanisms which are hidden below the surface of things, and not at all obvious or easily observed.

One explanation of a chemical change such as the burning of a piece of wood might be that the wood was made of a single homogeneous sort of matter, and that some of that substance changed its outward appearance to become smoke, while some more of it changed its appearance to become ash, and perhaps some more became heat or flame, though it still remained the same matter. The traditional scholastic view, based originally on Aristotle's theory of matter but much altered through the centuries, was of this kind. According to the Aristotelian view, any sample of matter might take on any Form—might for example become wet instead of dry or cold instead of hot—and so might change from being one of the four elements to another.

Another explanation might be that there are subsisting in inert matter various unseen and perhaps immaterial spirits or principles, rather like Aristotelian Forms, that cause it to show observable properties, and that chemical change involves one of these unseen spirits or essences replacing another, while the visible matter remains unchanged in substance, though changed in the way in which it presents itself to our senses. Originally the concept of the three Paracelsan or Spagyric principles, Salt, Sulphur, and Mercury (that is, the philosophical Salt, Sulphur, and Mercury, not the ordinary substances of those names) was of this kind. In either case, there was no need to suppose that the property of weight was unique in being preserved unchanged when other properties might change. However, laboratory chemists tend to forget philosophical niceties. In practice both these views of chemical change were modified and developed towards a third kind of explanation, though traces of them lingered on.

Such a third kind of explanation might be that the wood which is burnt was not homogeneous in the first place, but compounded of two or more components that are wholly material but essentially different in nature. Burning can then be explained as the separation of these components, one of them (or one group of them) forming the smoke, another the ash, and perhaps another the heat. In that case it would be natural to suppose that each component remained essentially the same during the burning, even if it changed its state of combination, and it would even be easier, though not necessary, to suppose that it retained the same weight. This concept of chemical change is much more natural for the chemist. In practice, it was convenient for him to suppose that the substances that he actually handled in the laboratory were either themselves simple or composed of simpler substances that could be extracted from them, and that the simpler substances retained their essential nature even when they entered into combination. A relic of the other two kinds of explanation lingered on in the assumption that the properties of a compound were a blend of the properties of its components, so that it would be possible to tell from the properties of the compound what the components were.

It is true that a large part of the antecedents of chemistry lay in alchemy. In the seventeenth century, indeed, that was not necessarily disreputable. The distinction that seems obvious to a modern chemist between his own scientific discipline and the pseudo-science of alchemy was not yet evident. As has already been mentioned, Isaac Newton, to name only one distinguished example, gave a large amount of his time to the study of alchemy, though he was perhaps a little old-fashioned and untypical of mathematical physicists in that respect. However, the alchemical approach was not at all helpful in looking for the kind of rational structure that the eighteenth century demanded. For one thing, alchemical writing is almost always in elaborate, obscure allegorical or metaphorical terms. It is often unclear whether the allegories and metaphors are to be interpreted as instructions for the purification of the soul and the attainment of a higher level of reality that was hidden from the senses and existed beyond the everyday world, or as recipes for material transmutations of base metals to noble metals. Since eighteenth-century chemists were quite sure that they did have to deal with the everyday world that is accessible to the senses, alchemical theories were not of great use to them, even though much knowledge of laboratory processes and of the properties of various substances might have been inherited from alchemists. Indeed, as we shall see, the alchemical symbols were still in use throughout the eighteenth century, chiefly for convenience rather than because they carried any occult meaning. Further, alchemy was certainly not quantitative, precise, definite, or rigorous—qualities that were increasingly expected in the eighteenth century.

In many ways two other antecedents of eighteenth-century chemistry were important. The medical tradition played a large part in the evolution of chemistry as an autonomous discipline, not only because physicians needed to prepare medicines and chemistry was therefore included in the curricula of medical schools, but also because many physicians used the skill and knowledge which they had acquired as part of their professional equipment to investigate chemical problems for their own sake or out of curiosity. The other essential contribution to the development of chemistry came from what we would now think of as industrial practice, though before the Industrial Revolution the term is misleading, and we should speak rather of the practical arts.

For some of such practical purposes it had to be assumed implicitly that weight was an unchanging property. Furthermore, it was natural to assume that the proportion of the simple substances to each other by weight in a given compound would always be the same in any sample of such a compound.

For instance, when a sample of an ore was assayed it was obviously assumed that a large quantity of it would have the same proportional composition as the sample, and that the weight of useful metal extracted from the large quantity would be in the same proportion to the weight of the ore as the weight of metal extracted by the same process from the sample assayed. Otherwise there would be no point in carrying out the assay. Similarly, apothecaries naturally assumed that if they weighed out the ingredients for a medicine in the prescribed quantities, the resulting medicine would have the same effects as the last time they weighed out the same ingredients in the same quantities. Otherwise the patient might die or at least fail to recover, which would tend to be bad for business.

Thus the traditional notions of the four Aristotelian elements and the three philosophical principles were modified in practice so that the elements and principles were understood by chemists to be distinct substances, which remained more or less unchanged during chemical reactions. Even after 1700 there was still some vagueness about the conception of them. It was only gradually that chemists acquired the habit of weighing all the substances entering into reactions or resulting from them, and of basing arguments on the assumption that the weight of each of them was conserved. However, the traditional elements and principles bore less and less relationship to the substances actually identified and named in laboratories. We find that although textbooks of chemistry may expound in introductory chapters a list of elements or principles derived from the Aristotelian Earth, Water, Air, and Fire, or from the Paracelsan Salt, Sulphur, and Mercury, often with the addition of Acid and Alkali, the substances described in the later, practical chapters are only distantly connected with the theoretical list. Presumably the chemist, as a representative of a branch of natural philosophy which was not yet accepted among practitioners of academically prestigious branches such as astronomy, retained the Aristotelian and Paracelsan notions in the introductory chapters because of a feeling that they were orthodox and respectable. As we shall see, the device of drawing up a table of chemical combinations, called a table of affinities or attractions, enabled him to give an empirical list of the substances actually encountered in the laboratory without immediately abandoning the orthodox notions or indulging in speculative theories of his own. Eventually, however, having ceased to be useful, the traditional notions of elements and principles ceased to be mentioned and the concept of elements derived from chemical practice replaced them.

The feeling that chemistry had not yet become a branch of natural philosophy, and that in order to attain the prestige of natural philosophy it ought to imitate the more illustrious branches such as astronomy or mechanics, also imposed a further obligation on the chemist. By the start of the eighteenth century those prestigious branches had established definite views of the methods which ought to be followed and the kind of explanation which ought to be accepted in natural philosophy. As has been mentioned above, and will be further discussed in the next section of this chapter, it was generally agreed among natural philosophers that speculation ought to be avoided and that all theories must be based on experiment or observation, if possible quantitative. Admittedly, there was much room for disagreement about what was a speculative system and what was based on sound empirical data. On the whole, philosophers tended to believe that their rivals' theories were rash and speculative and their own were not. However, the principle was accepted.

Also, it was generally felt that explanations ought to be mechanical, or if possible mathematical. The universe was treated as a vast machine. Natural philosophy was concerned with material things, not with the spiritual or immaterial. The visible workings of nature were supposed to be the results of the particular shapes, sizes, and motions of invisibly small particles. Chemists, who felt that their subject ought to be regarded as a branch of natural philosophy, and were conscious of a need to attain prestige by imitating the approach and methods of the more mathematical branches, therefore felt pressed to accept such mechanical explanations of chemical reactions and properties even though they did not help much in the solution of distinctively chemical problems and were unlikely to be experimentally verifiable. Also, a number of writers who were not chemists but mathematicians, or what would now be called physicists, helpfully expounded possible systems of mechanical explanations for chemical phenomena. Some of these will be described in the next chapter.

The belief in invisibly small particles of matter may have been helpful to chemists in some ways, as it provided them with a useful mental picture of the ways in which chemical reactions took place, and it implied that the particles would retain their characteristics, including their weight, during reactions. Also, it may have helped them to feel certain that a sample of a particular element would always remain a sample of that particular element and not change into another one, since its nature depended on the shape and size of its particles, and those characteristics would never change. However, it did not assist chemists very much otherwise. Indeed, it placed them in a dilemma. By definition, these particles were too small to be observed. Any theory about the relationship between their shapes and sizes and observable phenomena must therefore be entirely speculative. However, as we have seen, another important tenet of natural philosophy was that speculation should be avoided, and only theories derived by sound induction from reliable observations or experiments should be admitted. Nor were the traditional Aristotelian elements and the Spagyric principles much use in explaining chemical observations. Instead, chemists had gradually to find the confidence to evolve new kinds of explanation of their own. Thus in examining theories of chemical affinity and attraction we shall be examining the conceptual model of chemical composition and chemical change, different from the conceptual models of the mathematicians or physicists, which was developed by eighteenth-century chemists.

As soon as chemical change was seen as a matter of more or less stable and essentially different substances combining with each other—which is after all a very ancient interpretation, even though it did not become fashionable before the seventeenth century—it must have been observed that most substances were strongly selective in their reactions. That is, a given substance would combine with one substance in preference to another, or even separate from a substance with which it was already combined in order to combine with another. One of the most obvious tasks for a chemist was therefore to find out which substances combined with which. The next natural step for an enquiring mind was to look for an explanation of such tendencies. One explanation is represented by the commonest word for such a tendency, that is 'affinity'. Another is represented by the word 'attraction', which was introduced into chemistry through the influence of Newton. This term from the first implied that matter was particulate, and enabled chemists to try to make their subject quantitative by measuring the strength of chemical attraction. However, their attempts to do so were unsuccessful. Neither the mechanical philosophy nor the traditional theories of elements and principles were able to explain the selectiveness of chemical combination, or indeed various features which were important in chemistry. Chemical concepts such as those of affinity and attraction had to develop autonomously.

Before discussing these eighteenth-century concepts themselves, we must first look briefly at the ideas of the methods and concepts proper to natural philosophy that had been inherited from the previous century, at the background of national communities and industrial processes against which the emerging science of chemistry has to be seen, and then in Chapter 2 at the notions of chemical affinity and attraction which were held at the turn of the centuries by such men as Newton, Stahl, and Boerhaave.

Scientific assumptions in the Age of Reason and the Age of Enlightenment

European philosophers in the eighteenth century liked to think that their own time was one of enlightened reason, which had seen through the claims of superstition and could discuss man and nature in a spirit of calm and freedom from prejudice. They had a feeling for orderliness, symmetry, and elegance which is expressed in natural philosophy just as it is in literature and architecture. Their taste is seen, for example, in the music of J. S. Bach, or in the poetry of Alexander Pope or in the painting of Watteau or in the architecture of great houses from Sans Souci or Esterhazy to Kedleston Hall. By contrast, the seventeenth century often seems to have been a time of violence and extremism, when religion was a matter for wars rather than for discussion. Nevertheless, there is a continuity in the development of scientific attitudes and methods between the two periods. Indeed, it was only a privileged few who could seek order and elegance in the eighteenth century: for the great majority there was as much poverty and misery as ever. Yet it was the privileged few who produced modern science. I shall not attempt here to give a complete history of scientific methods, let alone a complete history of philosophy in the periods concerned, nor to discuss in detail the philosophical problems raised by the ideas which I shall mention; but rather to draw the reader's attention to certain topics that are especially relevant to the subject of this book.

Philosophers in the early seventeenth century were in rebellion against the Scholastic system, taught largely though a little unfairly in the name of Aristotle, which had been dominant in the universities of the later Middle Ages. Francis Bacon, for instance, criticized the reliance on authority and on deductive logic of the Aristotelian system and advocated induction. He emphasized the necessity for experiments, and laid down rules for eliminating circumstances which were not the cause of the phenomenon in question, and hence for establishing its true cause.[11]

Galileo, though working within the framework of Renaissance Aristotelianism, used a new, mathematical approach to mechanics which replaced the traditional mechanics based on Aristotle, appealed particularly to experiment as the proper test of theories, and distinguished between the primary and secondary qualities of things—that is, between such qualities as number, shape, size, and position, which cannot be separated from the bodies to which they belong, and those qualities which are merely the effects on our senses of the primary qualities. The primary qualities were those which could be expressed numerically and treated mathematically.[12]

Descartes, looking for a new foundation of certainty to replace the foundations laid by Aristotle, found that it was possible to doubt his knowledge of everything except the existence of his own thoughts, and consequently of his own mind, and the existence of God. On this basis he proceeded to erect a complete new system of philosophy to replace the old. It implied inevitably that there was a sharp distinction between mind and matter, and that we can obtain direct information about the world of matter only through our senses. Matter was defined as that which had extension, and again the primary qualities were seen as the only ones which were real. Descartes believed that it was possible in theory to deduce all of science from first principles. However, having shown from those principles that the evidence of the senses was to be trusted, since God would not set out to deceive us, he considered that in practice the most convenient way of discovering the facts of the material world was by the use of the senses, that is through observation and experiment. Descartes, however, saw experiment as a way of finding which of several hypotheses suggested by the mind was to be preferred, rather than as the source of material on which to build hypotheses.[13] Later it became the orthodox view that possible theories should not be built by pure speculation, but should be reached by generalization from actual observations.

During the seventeenth century, then, it came to be generally accepted among natural philosophers that the basis of their studies of nature must be observation and experiment, and that the ideal scientific theory was a mathematical treatment of primary qualities, using data derived solely from observation or experiment. The type of explanation favoured was the mechanical explanation—that is, an explanation of observable phenomena in terms of materialist mechanisms which could be pictured as similar to the observable mechanisms of everyday devices, which could function without any immaterial or mysterious agencies, and which would follow determinate and discoverable laws. Chemists in the eighteenth century often criticized their colleagues for building 'systems', that is theories which were not founded on observation and experiment but were devised by the use of unsupported reason. Indeed, the Cartesian system itself was often viewed as an example of a system constructed entirely by reasoning and not securely based on experiment.

During the seventeenth century there were instituted a number of formal or semi-formal societies of natural philosophers whose main activity was doing or discussing experiments. Such were the Accademia dei Lincei in Rome and the Accademia del Cimento in Florence, on a more formal basis the Royal Society in London and the Académie Royale des Sciences in Paris, and the informal associations which were the forerunners of the last two. The motto of the Royal Society of London was and is 'Nullius in Verba', implying that members should trust only the evidence of their own senses.

Associated with the insistence on mechanical explanations and on the primary qualities of matter as the objective reality with which natural philosophy should deal, as opposed to the Forms of Aristotelian philosophy, often now given the epithet 'occult', was a belief that matter was essentially particulate. The primary qualities were then reduced to the shape, size, and motion of the particles of which matter was composed. The behaviour of these particles could be suggested as the mechanism causing phenomena that could be observed, even though the particles themselves were insensibly small. Mechanical explanations of that kind, often capable of being expressed mathematically, stemmed from the ancient Greek atomic theory, particularly as expounded in Lucretius's *De rerum natura*, which had begun to be read again in the early seventeenth century. Gassendi's writings were largely responsible, though as a Christian he naturally differed from Lucretius over the nature of God. According to the ancient atomic theory there is nothing in the universe except atoms and empty space, with the possible exception of the gods, who according to Epicurus exist in a distant part of the universe remote from human affairs. The human soul and spirit are composed of atoms in just the same way as everything else, and all that happens is due merely to the movement and arrangement of atoms. The atoms themselves are indestructible, but their arrangement—and so the objects that they form by clustering together—is always subject to change.

Descartes also believed that matter was particulate; but his theory was greatly different from the atomic theory. According to him, a vacuum was an impossibility, since whatever had extension was by definition matter, and all space was filled with one or other of the three kinds of matter, differing in the sizes of their particles, and corresponding roughly in their properties with the Fire, Air, and Earth of Aristotle. The first kind, consisting of the smallest particles, was also called ether. The larger particles were not atomic, but divisible, so that they could be worn down into the smaller ones. The reason why the particles of the third kind, the largest, held together in a mass, so forming a body with substance, was not any attraction between them, nor external pressure, but simply the absence of relative motion. In Descartes's system, all forces or pressures must be transmitted by the particles of one of the three kinds of matter, in contact with one another; and action at a distance was inconceivable. In particular, gravitation was accounted for by the action of vortices in the ether.[14] That naturally became a point at issue between the Cartesians and Newton.

Newton's most influential achievement was that of producing an elegant mathematical system, based on his laws of motion and law of gravitation, from which in principle all the movements of the sun, moon, and stars could be determined. That became for the eighteenth century a model of what science ought to be like—neat, simple, materialist, orderly, logical, and mathematical. However, Newton's attribution to the particles of matter of powers of attraction and repulsion (which will be discussed in more detail in Chapter 2) was also extremely influential.

The ideas of the natural philosophers were in harmony with those of the epistemologists and metaphysicians. The modern separation between the natural sciences and the humanities was unknown to late seventeenth-century Europe, and philosophers in the modern sense were frequently interested in natural philosophy. John Locke was such a philosopher. Although he is probably best known (outside Oxford) for his work in political philosophy, he was deeply interested in natural philosophy, and knew Robert Boyle and John Wilkins during his time in Oxford.

Locke argued that the ideas in the mind (which he treated almost as if they were atoms in a box) could be produced only by sensation or by reflection on other ideas. His account of them has much of the flavour of the mechanical explanations of the natural philosophers. He emphasized the distinction between primary and secondary qualities.[15] The chief difficulty in Locke's epistemology is to explain how ideas formed by a single sense-impression of an individual object could be associated by what Locke calls reflection in such a way as to form general ideas or ideas of classes of objects. However, the point for our present purpose is the implication that the only possible source of information about the natural world was sensation.

Berkeley on the other hand denied that primary qualities could be distinguished from secondary qualities as less dependent on perception, and argued that existence meant no more than being perceived by a mind.[16] Condillac, who familiarized France with a philosophy similar to Locke's, had therefore to produce lengthy arguments in order to maintain the existence of anything external to the mind at all.[17]

Philosophers also made use of the atomic theory. Thomas Hobbes, another philosopher who was deeply interested in natural philosophy and indeed in pure geometry, founded his moral and political philosophy on the belief that human beings are no more than machines made of particles, so that all human thoughts and feelings are to be interpreted merely as the motions and interactions of the particles.[18] Hence he argues that what is good is simply what produces pleasure, and what is bad is what produces pain: no loftier meaning can be attached to good or bad. Like Locke, Hobbes is probably best known for his political philosophy. Although he suffered fierce attacks for atheism and the destruction of orthodox ethics, in *Leviathan* he founded a justification of absolute monarchy on his theory of human nature and the origin of human societies. In this he was opposed to Locke's view that society was based on a social contract. However, he also engaged in vigorous controversy with Robert Boyle on the proper basis for scientific theory. As well as questioning the validity and reliability of Boyle's experiments with his air pumps, which tended to imply the possibility of a vacuum, Hobbes insisted that the proper way to produce a philosophical theory was by reason. He argued that the experimental philosophy of Boyle and his colleagues in the Royal Society, which accepted experimental results, well witnessed and capable of replication, as the only source of information about the natural world, was entirely wrong. However, Boyle's view prevailed in the next century.

The mechanical interpretation of all phenomena, including the human body and mind, in the manner of Hobbes, which is presumably derived from Epicurus by way of Lucretius and the accounts in Diogenes Laertius and Aristotle, persisted into the eighteenth century. David Hartley, for instance, based his account of sensation, and of the formation of passions, affections, and a moral sense from ideas received through sensation, on the vibrations of the white medullary substance of the brain, reproducing the vibrations of the objects of sensation, though he avoided committing himself to complete materialism.[19] The attraction and repulsion of particles were also brought into moral philosophy, for instance by the Unitarian theologian Joseph Priestley, also one of the foremost natural philosophers and notorious as a political pamphleteer, in his *Disquisition on matter and spirit*.

Priestley was arguing that the soul was wholly material, though immortal; and he quoted from Locke, Newton, and the Jesuit natural philosopher Boscovich (who was naturally not pleased to find his work used in such a context). Priestley then wrote

> All the properties that have hitherto been attributed to matter, may be comprised under those of *attraction* and repulsion (all the effects of which have been shown to be produced by powers, independent of all solidity) and of *extension*, by means of which matter occupies a certain portion of space. Besides these properties, man is possessed of the powers of *sensation* or *perception*, and *thought*.[20]

However, those powers could also be accounted for by the same three properties of matter, attraction, repulsion, and extension; and since causes and kinds of substances should not be 'multiplied without necessity', there was no need to attribute man's mental or spiritual activities to any immaterial principle.

Such, then, was the philosophy of matter which British followers of Newton adopted. In France, however, and other Continental countries, the influence of Descartes prevailed even among chemists well into the eighteenth century. There philosophers were more inclined early in the eighteenth century to construct systems *a priori*, whereas Newtonians set out to produce only those theories which were required by the experimentally observed facts, and not to 'feign hypotheses'. Even in France, however, it was a serious criticism of a chemist in the mid-eighteenth century to claim that he had constructed a speculative system. There were other points of difference. Descartes had defined matter as that which has extension, and so thought vacuum impossible, space which was not occupied by the particles of the two larger kinds being filled with the very fine particles of the first kind of matter, which he called the ether. It was supposed by Cartesians that one particle could affect another particle only if the two were touching, since no other effects could be conceived in terms of the primary properties of matter.

On the other hand, Newton's influence through the *Principia mathematica* was against assuming any such simple mechanical explanation. His method in that book was to show that from a simple mathematical law such as the inverse square law of gravitation, together with the laws of motion, all the movements of the planets and of terrestrial bodies could be predicted; and although at times he too discussed mechanical explanations in terms of pressure transmitted by an ether, the best-known parts of his work avoided such concepts. His followers at the level of the mathematical kind of natural philosophy therefore spoke instead of a force of attraction operating between particles which were at a distance from each other. To the Cartesian, however, attraction came under the heading of occult powers, mysterious and irrational properties which were best rejected in favour of clear pictures of the interaction of particles in direct contact. The Cartesians therefore objected to Newtonian physics as based on occult powers, and the Newtonians objected to Cartesian ideas as speculative and not properly based on experiment.

Cartesian philosophy remained dominant in France until the 1730s. Among the generation then new, Newtonian physics came to be generally accepted, largely owing to the advocacy of such authors as Voltaire.[21] From then on the sceptical and rationalist spirit of the French *philosophes* found the outlook and methods of Newton's *Principia* especially congenial. Several philosophers made use of the notions of attraction and repulsion. Baron d'Holbach, for example, a devoted atheist and materialist who translated several German chemical books into French and was one of the Encyclopedists, drew an analogy between the principles of desire and aversion which govern the human machine and the principles of attraction and repulsion which are supreme in the mechanism of Nature.[22] Repulsion and attraction were also important in the natural philosophy of Immanuel Kant, perhaps the greatest of eighteenth-century philosophers.[23] Kant, however, cracked the complete confidence of the majority of eighteenth-century thinkers in the power of reason to solve all problems by denying that the human mind was competent to discover the true nature of the universe.

Although eighteenth-century chemists must have been aware of the use which was made of Newtonian physics by speculative philosophers, nevertheless the striking thing about their work is how little they referred to mechanical explanations in terms of the properties of particles. They were indeed deeply imbued with Newtonian concepts of scientific method. They often insisted that only experimental evidence was to be accepted, and that hypotheses must not be considered except when they were properly derived from experimental evidence; and they agreed that mechanical explanations were rational and desirable. Also, they greatly admired the order and power of Newtonian physics. Nevertheless, because it seemed to be almost impossible by definition to get any experimental evidence about the size, shape, and motion of the minute particles of which they virtually all assumed matter to be composed, they seldom permitted themselves to speculate about the mechanisms by which they undoubtedly believed those particles to produce the phenomena which they could observe.

Also, it became fashionable among natural philosophers in the eighteenth century to frame explanations not so much in terms of the primary properties of particles but in terms of what have come to be called imponderable fluids. Such fluids—light, the matter of heat, the electric fluid or fluids, and the magnetic fluid, for example—were invisible, generally assumed to be weightless, odourless, impalpable, and in fact devoid of all properties except the one which made each what it was, and the properties of attraction and repulsion attributed to the particles of which they were supposed to consist. For example, apart from attraction and repulsion, the only property of the fluid of light was that it produced vision, and the only property of the matter of heat was that it produced the sensation of heat. Attraction and repulsion were also used to explain some of the incidental properties of each fluid, such as the attraction between the two different electric fluids and the repulsion between samples of the same fluid, that is between objects with like charges. However, although chemists accepted the existence of these fluids and made use of them—indeed, phlogiston, a central feature of eighteenth-century chemistry, was itself a kind of imponderable fluid—they were usually very cautious about speculating on the invisible mechanisms by which such fluids worked.

As well as mechanical explanations, there were other features of the more respectable branches of natural philosophy which chemists admired. The paradigms of success were astronomy and Newtonian mechanics. They were precise, quantitative, and mathematical, and had ancient Greek antecedents. Astronomy had indeed long been associated with astrology, which by the start of the eighteenth century was hardly regarded as a proper part of natural philosophy; but the association could conveniently be forgotten. More important, astronomy and mechanics seemed, from observation rather than from *a priori* theories, to have evolved clear, definite, and quantitative laws from which verifiable predictions could be made. Chemists who aspired to intellectual respectability therefore felt the need to try to imitate those features.

Another factor was the Enlightenment urge to impose on nature order and pattern, which might be expressed in systems of classification. Mechanics could be displayed, like geometry, in the form of numbered theorems; stars could be catalogued; tables could be drawn up of the motions of the planets like mathematical tables; but chemistry seemed (as it still seems to many A-level and even first-year university students) a mass of random recipes and descriptions. Nowadays the Periodic Table is supposed to supply a framework which relates the properties of substances to a theoretical structure. Eighteenth-century chemists were keenly aware of the need for some such framework. However, the traditions of seventeenth-century chemistry did little to supply it. As we shall see, tables of affinity or attraction were in part an attempt to fulfil this need.

It has already been pointed out that chemistry was hardly regarded as a serious branch of natural philosophy before the end of the seventeenth century. It was far from being a tidy, organized, rational discipline, in which all the phenomena could be deduced from a few clear principles, or in which the phenomena could be explained simply by reference to the primary properties of particles. About its nomenclature and its theoretical concepts there still clung as much of the reek of alchemy as of Aristotelianism, and the chemical laboratory was still full of dark mysteries for the uninitiated. Neither assayers of ores nor apothecaries, who did much of the work in chemistry, came very high up the social or academic scale. It is true that even more was done for chemistry by physicians, who were higher in social and educational status; but even so they had not succeeded in making chemistry a respectable science until the close of the seventeenth century. Much of the credit for establishing chemistry as a branch of serious learning must go to Robert Boyle, whose wealth and secure social position as a son of the Earl of Cork were a sound basis for academic respectability.

Boyle regarded himself as a philosopher, with the secure status of the astronomer or geometer rather than the quite different situation of a chemist. Indeed, he spoke of 'mere chemists' in the third person. Nevertheless, he set out to show that all phenomena, including those of chemistry, could be explained through the sizes, shapes, and motion of the minute corpuscles of which all matter was composed, and inveighed against the Spagyric notion that the three principles (Salt, Sulphur, and Mercury) were the constituents of matter, and against the Aristotelian view of the four elements, as being irrational and contrary to the experimental evidence.[24] For example, such observable properties as the texture of a substance were due to the shape and arrangement of the particles which composed it. However, Boyle's own writing was too diffuse and too unsystematic for his influence to have been as great as might have been expected. (Shapin and Schaffer have suggested that

Boyle's prolix, convoluted, and unadorned style was to show that he was a modest and reliable reporter of experimental results.) The seventeenth-century author who had by far the strongest influence on eighteenth-century chemistry was Newton himself, whose scanty published writing on chemistry proved extremely suggestive. The ideas of Boyle and Newton on chemical affinity and attraction will be discussed in Chapter 2.

The environment of chemistry

Not long ago there used to be a division among historians of science between externalists and internalists—that is, between those who emphasized the influence on the development of scientific ideas of external factors, such as economic, social, and institutional pressures, and those who concentrated entirely on the working out of new scientific ideas in accordance with their own internal logic. It is surely evident that ideally the history of science should be a synthesis of the two. The internal history ought to be seen as part of the whole complex involving economic, social, and cultural changes, and as inseparably interwoven with them. The difficulty is that there is not often, if ever, a simple relationship of cause and effect between a single factor in the external context of science and a particular internal development. Rather, we have to see the various external factors as part of the whole, and as determining together with the internal tensions of science at a particular time the range of possibilities for development, from which the practitioners of science, themselves the products of the conditions of their time, select for various reasons their particular course of thought and experiment by which they make actual in their own ways some of its potentialities.

In this book I am concerned mainly with the internal development of chemistry. However, plainly that cannot be properly understood without some description of its environment. In the eighteenth century there was no such thing as a scientific profession, in the sense of a group of people who not only earned a living by science but consciously formed a community with a special vocabulary, training, organization, and ethos. Many natural philosophers were amateurs, often physicians or parsons. There were indeed in many countries teaching posts which might well be occupied by men who carried on scientific research; and there were even a small but increasing number of posts associated with industrial practice. Much of the impulse towards chemical research came from the interest of the subject itself, as with any other branch of learning. However, it was expected that chemistry, like other branches of learning, could be of practical use, both in medicine and in manufacturing industry. Part of the reason for encouraging the teaching and practice of chemistry was therefore utilitarian, though as a matter of fact the extent to which theoretical chemistry actually was of practical use was limited. On the other hand, the industrial arts often provided information which was valuable material for the theoretical chemist. It must also be appreciated that there were great differences between different countries in the amount of institutional support for chemistry.

Chemistry had been taught in European universities well before 1700, though almost solely for the benefit of medical students. At Leiden, for instance, Franciscus Sylvius (Dubois or de le Boë) became professor of medicine in 1658, and since he firmly believed in chemical explanations of physiological processes he included chemistry as well as standard medical topics in his lectures. He learnt much from the writings of Van Helmont. As well as van Maets and Le Mort, the great teacher Hermann Boerhaave was among his successors. Another influential teacher was Johann Conrad Barchusen at Utrecht.[25]

It was normal for the capital town of each of the numerous German states to have its own university if it could, and a great deal of chemical research was done by the professors of such universities during the eighteenth century. These German universities differed very much, as did the German states, in their religious affiliations. Halle, for instance, where Stahl taught, was a centre of Pietism. Much of the chemical research done in them had a bearing on medicine; but there was also a considerable interest in metallurgy. The patronage of the ruling families of the German states was particularly important to professors in these universities, and it is evident that in a time of mercantilism and enlightened monarchs the mining interests of several German states made the relevant parts of chemistry especially significant in the work of German chemists.

Sweden, though larger than most of the German states, also had university teaching which was directly encouraged by the king, and teachers who concerned themselves with metallurgy and other topics of practical importance to the state. Natural philosophers in German states, however, tended to feel that they owed intellectual allegiance to Paris.

France, in contrast with its neighbours to the north and east, had a strong centralized administration. It had a far larger population than most other European countries, only the much less tightly organized Austrian Empire being comparable. The peasant class, though fairly prosperous, was heavily taxed to support the large privileged class of nobility and clergy, and there was also a comparatively large middle class of merchants, manufacturers, tax-farmers, and bankers. These wealthier classes were therefore able to produce a society of great brilliance, which led Europe in the arts and fashions as well as in philosophy and learning. Many of the provincial capitals such as Bordeaux and Dijon must have had the better of any comparison with the German provincial capitals; but nobody in Paris, the centre of the administration, was ever in any doubt that there was the capital of Europe in everything that mattered.

The chief event in eighteenth-century France was obviously the Revolution, and the same spirit of sceptical and enquiring rationalism that prevailed in French natural philosophy also led to the enunciation of the Rights of Man and the political philosophy of the Revolution. The King's Government had realized that there were political dangers in freedom of thought, and there was a Royal censorship before the Revolution. However, the Revolution and the events that followed did not apparently cause a discontinuity in the development of chemical theory, even though in the long run no doubt they had a considerable indirect effect, and although they radically changed the lives of a few chemists such as Fourcroy. Indeed, they ended Lavoisier's.

The industrial revolution in Britain was to reveal a certain backwardness in French industrial methods, particularly in such things as the use of coal; but that can hardly have been realized in France before the end of the eighteenth century. The Revolutionary government made remarkable efforts to galvanize French industry into producing the necessary supplies for the armies. Among the movements that were fashionable in mid-century Paris, however suspect it may have been to the royal government, was that of the *philosophes*, and a polite interest in natural philosophy was very much in fashion. Although there was less social mobility between the classes in France than in Britain or some other countries, a reputation as a scientist might help a man to rise from the bottom of the middle classes to an enviable position. Royal patronage of science was particularly effective in France, for the members of the *Académie Royale des Sciences* were paid a small salary, and expected to serve in return on various commissions which were set up from time to time to advise on practical questions.

Chemistry had long been taught publicly by a series of distinguished professors (under the direct patronage of the king) at the *Jardin du Roi*. Also, various industrial posts were available for scientists—for instance, Macquer, who wrote about dyeing as well as pure chemistry, was superintendent of the porcelain works at Sèvres, inspector of dyeworks, and director of the Gobelins. *Le cumul*, the accumulation of several paid jobs by one man, is an old French tradition. Guyton de Morveau was a director of the saltpetre works before establishing his soda factory, and Berthollet, who provided a theoretical basis for the practice of dyeing, succeeded Macquer in 1784 as inspector of dyeworks and director of the Gobelins. Thus it was possible to make a living as a chemist, at least for some people, even in eighteenth-century France. In general, Paris can be regarded as the headquarters of eighteenth-century chemistry.

Britain, on the other hand, with much less central control, a small though vigorous nobility which was often not averse from patronizing philosophy or even profitable manufactures, and a large middle class, but a great deal of social mobility between the class of artisans and yeomen and the middle class, left much more to private enterprise. Membership of the Royal Society brought no salary. Natural philosophy was rather a hobby for gentlemen and a subject for interesting discussion than a matter of state concern. Patronage, of course, was important to those who had no other source of wealth; but it was the patronage of the nobility rather than centralized royal patronage. No doubt there was some money to be made by writing successful books, or books which would be well rewarded by a noble patron. Some private individuals, such as Bryan Higgins and Peter Shaw, offered courses of lectures by way of private enterprise; and at a rather lower level in the social scale there were professional lecturers such as John Warltire who toured the provinces giving courses of lectures. However, there were not many in Britain who made a living from chemistry.

There was, indeed, some teaching of chemistry in the universities. At Glasgow and Edinburgh, where the medical schools owed a great debt to Boerhaave, chemistry was established by Plummer, Cullen, and Black; and it is known that many who were not medical students, and some who were not university students at all, attended Black's lectures. There was also some teaching of chemistry at Oxford and Cambridge, though not for examination purposes.

However, although the less rigidly controlled social structure of England and Scotland evidently gave less organized support to scientists than that of France, it also helped to produce the vigorous commercial and entrepreneurial spirit which was among the most important of the many factors bringing about the industrial revolution here. In France, scientists might advise and be appointed to direct some of the *manufactures royales*, and a few might set up in business for themselves: in Britain, everybody seemed to be setting up in business for himself and looking for new methods of improving agriculture and manufactures. The Nonconformists, barred in England from public office as an outlet for energy and ambition, were prominent among the innovators and entrepreneurs; but they were not the only ones. There were provincial societies, formal and informal, for the study of such subjects as natural philosophy in the latter part of the eighteenth century in both France and Britain, for example the societies of Dijon, Rouen, and Arcueil in France, and the Lunar Society of Birmingham and the Manchester Literary and Philosophical Society in England; and such societies are an important part of the background of research and scientific communication in both countries. However, the contrast between the comparative centralization of France and the independence of British natural philosophers is still important.

The industrial background

To understand the background of eighteenth-century chemistry we must also remember the considerable growth of the chemical industry. Although the contribution of the practical arts to what was later called pure science was certainly greater in this period than the contribution of pure science to industrial practice, nevertheless it was normal for natural philosophers to claim that much of the value of their work lay in the help which it ought to be able to give to the engineer and artisan, on whom the increase of wealth depended. The founders of the Royal Society, for example, had the practical motive in mind, as had their patron Charles II. Quite apart from the medical uses of chemistry, which were always considered important, a great many chemists were interested in industrial applications.

Many of the chemical arts were, of course, ancient or at least very old—such as the extraction of the traditional metals from their ores, dyeing, glass-making, soap-making, the extraction of soda from ashes or kelp, the preparation of coloured glazes for pottery, lime-burning, distilling spirits from wine, cider, or other sources of alcohol, and extracting tar, pitch, and resin from wood. Mineral acids were generally made on the spot by the artisan as he needed them. Thus a wide range of simple chemical processes was known on the small industrial scale. However, such processes were generally carried on by craftsmen or artisans working by rule of thumb rather than on any theoretical basis, and the products were either extracted from naturally occurring raw materials or produced by simple combinations of natural materials rather than by methods based on chemical theory. The contribution of scholars in the fifteenth and sixteenth centuries had been the searching of ancient writings in an attempt to discover lost processes, more often than the inventing of new processes.

During the seventeenth century, however, educated men did produce some innovations in chemical manufacture. Cornelius Drebbel, for instance, a Dutch immigrant to England, devised new weapons for the Navy, an improved method of producing sulphuric acid, devices for controlling the temperature of furnaces, and new dyeing processes. Charles XI of Sweden had a laboratory built where soils and ores were analysed with a view to finding uses for their products. Sir Nicholas Crisp instituted a new method of brick-making and proposed a scheme for the manufacture of copperas. Many books describing the best practice in the chemical arts and crafts from various parts of Europe were published. Nevertheless, many mcre suggestions were made by learned gentlemen than were actually adopted in the industry.

From the middle of the eighteenth century the developments known as the Industrial Revolution enormously increased the demand for basic chemicals to be used in the essential processes of other industries—sulphuric acid, soda, and later chlorine, for instance. The enlargement of the scale of chemical production to meet this demand was brought about by new methods and a new outlook. The entrepreneurs were no longer merely artisans, but were often educated men of standing. Frequently the problem was not to find a source for the product, but to replace an old method of making it on a small scale by a method of making it in larger quantities, perhaps continuously, from much the same raw materials. It is true that one of the most important innovations, the use of coal or coke in furnaces instead of wood, and to produce by-products such as coal tar, was probably a matter of trial and error rather than of scientific method. Nevertheless, some of the innovations of the eighteenth-century chemical industry were introduced by men who knew a great deal about laboratory and theoretical chemistry; and undoubtedly the search for better ways of carrying out industrial processes was an important stimulus to laboratory chemistry.

Sulphuric acid, or rather in eighteenth-century terminology vitriolic acid, was used in a wide variety of processes, for instance in pickling and cleaning metals, in bleaching and dyeing textiles, and towards the end of the century in the production of soda. The method of producing vitriolic acid *per campanam*, that is by burning sulphur with saltpetre (potassium nitrate), as Drebbel had suggested, under bell-shaped vessels of glass and earthenware, seems to have been introduced into England about 1720.[26] However, the transition from the laboratory scale to the small factory scale was not made until Ward and White began to make vitriolic acid *per campanam* at Twickenham in 1736.[27] They used glass vessels with a capacity of forty or fifty gallons, and so greatly reduced the price of the acid to their customers, to two shillings a pound.

The process was satisfactory chemically, but the quantity of acid which could be manufactured was limited. It was impossible to make glass vessels larger than those used by Ward and White, and the obvious difficulty in using metal containers was that metals are generally attacked by vitriolic acid. The difficulty was solved by John Roebuck, in partnership with Samuel Garbett, by the introduction of the lead chamber process.[28] Roebuck, a graduate of Leiden University and a very active entrepreneur, made use of the fact that lead is not attacked by vitriolic acid, and made his chambers for producing acid out of lead. The process spread rapidly over the country, and was also introduced into Holland, France, and Germany. Because of the difficulty of transporting the acid, it was common for a manufacturer to set up his own lead chamber plant to produce the acid needed for other processes on the site.

The rewards awaiting any manufacturer who could find a better means of making alkali than extracting it from ash were obvious. The process which eventually succeeded was devised by genuinely rational and scientific methods, as were several earlier attempts; and several of those concerned in those attempts were competent or even distinguished chemists. Here again there was no difficulty over a source of raw material. All processes started with common salt. The problem was to convert the salt to soda, and get rid of the unwanted by-products cheaply and efficiently.

For example, Joseph Black, whose doctoral dissertation had first made clear the relationship between mild and caustic alkalis, was associated with Roebuck and with James Watt in experimental syntheses of alkali which they would have put into production on the industrial scale if the duty on salt, their main raw material, had not been so high.[29] Another associate was James Keir, who later set up an alkali and soap works near Dudley, became a member of the Lunar Society, and translated Macquer's chemical dictionary. In 1771 the group opposed an application by Alexander and James Fordyce, then of London, for a patent for manufacturing soda from salt. Evidence was given that Keir had made soda artificially. Several other patents for the manufacture of alkali were taken out in England during the next decade.

An important contribution to the foundation of the British alkali industry was made by Lord Dundonald, a man of wide scientific interests. With Lord Dundas he set up an alkali works at Dalmuir in 1795. However, the works which he set up in 1796 at Walker upon Tyne in partnership with William Losh and others is better known.[30] They used vitriolic acid, like most soda manufacturers of the period, to convert common salt into Glauber's salt (sodium sulphate), and then converted it into sodium sulphide, from which either mild or caustic soda could be produced. Several other factories followed in the nineteenth century. However, the greatest development had to wait until the Leblanc process was introduced, and until Muspratt's enterprises in the Liverpool area were able to benefit from the repeal of the duty on salt in 1823.

Meanwhile, however, the Leblanc process had been developed in France.[31] The need for artificial soda to replace or supplement supplies of soda extracted from ashes was equally great there, especially when the entry of France into the American War of Independence brought about restrictions in imports, and later when the Revolutionary wars not only produced further restrictions on imports but also increased the demand for potash (for making saltpetre for use in gunpowder) and consequently the demand for soda as an alternative. The *Académie Royale* had offered a prize for a process for manufacturing soda in 1781, and it was offered again in 1786 and 1788. However, it was never awarded, largely because the grant of *privilèges* for the exploitation of various processes superseded it. Grants were made to Malherbe and Athénas in 1780 and 1782 for a process in which common salt was converted to the sulphide by treatment with vitriolic acid and then fusion with charcoal, just as in the laboratory process published by Duhamel in 1736.[32] The original feature of the method was the addition of scrap iron or particular iron ores to the fused mass. Soda could then be dissolved out.

Grants of *privilèges* were also made to Guyton de Morveau in 1782 for an entirely different process, depending on the emergence of soda from a paste of lime and brine in damp air, and to Hellenweger in 1783 for a process which started like Malherbe's by converting salt to sodium sulphide and then used various methods for extracting soda. However, these three grants were revoked in 1788 in favour of a new grant of a *privilège* to Guyton and Bullion for a number of different processes; and in the next year they took into their partnership Carny and Géraud de Fontmartin, who had also applied for *privilèges* for several processes.

None of these attempts produced successful manufacturing plants. As is well known, the process which was eventually successful was that which as mentioned above was developed by Leblanc, although because of financial trouble after the confiscation of the estates of his backer the Duke of Orléans the plant which Leblanc set up never quite went into full production and the process was eventually exploited by others. The distinctive feature of the Leblanc process was the addition of chalk as well as charcoal to the sodium sulphate to give a mixture which on fusion produced mild soda. The soda could be leached out from the mass of 'black ash', leaving calcium sulphate.

Some historians have described the work of the men who devised these processes as if it were mere guesswork, or trial and error which succeeded by some lucky chance. That seems quite implausible. It is true that natural philosophy was seldom of much use in such practical fields as mechanical or civil engineering, even though the main motive for the popularity of natural philosophy was precisely the hope that it might be of practical use. In the eighteenth century the chief reason why men with a knowledge of science were successful in industrial innovation was that they brought to technical development the rational, empirical outlook of the Enlightenment, and that they used systematic methods of experimentation, rather than that scientific theory was much help to them.

Nevertheless, the chemical industry seems to have been exceptional. Virtually all of those whom we have mentioned as projectors of soda manufacturing processes, for example, were men of education, with a knowledge of laboratory chemistry, at a time when philosophical chemistry was fashionable. It is hard to believe that they did not devise their processes as the result of laboratory experiments guided by chemical theory to some extent, even if much of the theory was crude and misleading and the insight which it gave into the chemical reactions involved was far from thorough. Plainly the difficulty in soda manufacture was finding a way of converting sodium sulphate to soda. A knowledge of the relative strength of the chemical affinity of various substances for each other, even if it was superficial and over-simplified in comparison with twentieth-century concepts, would be an obvious guide to possible methods of achieving the aim. Possibly the published work on chemical affinity and attraction was not after all of very much practical use when it came to the point; but at least there must have seemed to be a good chance that it would help in planning industrial processes.

The only acknowledged example of its being of use is apparently that of Fourcroy, who in trying to extract the copper from bell metal for making guns for the revolutionary armies used his knowledge that tin has a greater affinity than copper for oxygen.[33] However, chemists who were developing industrial processes naturally kept their ideas secret as far as possible, and so there may be other examples which were not recorded. The point is that the possibility of industrial uses of such ostensibly pure chemistry was known and understood, and must often have been in the minds of those experimenting and writing on the subject. On the other hand, the chemical industry produced a vast amount of practical observations which provided the raw material for the theoretical chemists. In the end, the most important development in chemistry in the eighteenth century was simply the enormous increase in the number of different substances described and distinguished from each other, in the number of their known properties, and in the number of different effects and reactions observed and recorded.

The dilemma of chemistry

To sum up, then, the prospect before early eighteenth-century chemists: their chief needs were to develop concepts and kinds of explanation which would help them to feel that they understood chemical processes and to predict the results of reactions which had not previously been observed. They also needed to impose order on the enormous quantities of substances which had been distinguished and of observations of their properties, presumably by fitting them into some logical classification. As practitioners of an activity which had hardly completed its progress towards acceptance as an academically and socially respectable branch of philosophy, they felt obliged to try to imitate senior branches which were seen as free from occult beliefs, rational, orderly, based on experiment and observation, quantitative, and summarized in simple mathematical laws from which the future movements of any object in the universe could in principle be predicted if its present position and velocity and the forces acting on it were known. Since such laws must be found by induction from experiment, it was not acceptable to deduce those which applied to chemistry from a philosophical system constructed *a priori*. Instead of such obsolete methods, the hope must be that when the mass of observations was set out in a clear and systematic tabulation, not only would it be easier to make sense of them, but also patterns would be seen from which the required predictive laws could be inferred.

At the start of the eighteenth century, and indeed for some time after that, chemists felt obliged to use the rather vague notions of principles or elements offered them by traditional philosophy, that is the Aristotelian theory of the four elements and the theory of the three principles, even though those notions did not really suit the needs of chemists or help to explain the actual phenomena observed in chemical laboratories. During the eighteenth century these traditional notions were gradually modified, as we shall see, for example by making elements such as earth and principles such as salt into categories including several substances. At the same time, for ordinary purposes chemists drew up lists of substances which were actually treated as elementary in laboratory practice. In the end, this unofficial list became, so to speak, official in the work of Lavoisier.

Furthermore, chemists felt obliged to respect the mechanical explanations, using the shapes and sizes of the invisibly small particles of which matter was supposed to consist, that were favoured by natural philosophers. This view of the nature of matter was helpful to chemists to a certain extent, inasmuch as it helped them to form a satisfying mental picture of the unseen mechanisms underlying chemical reactions, and probably helped them to decide that the weight of each substance would be unchanged during a chemical reaction. Yet by definition it seemed impossible to obtain any experimental evidence about invisibly small particles, and most chemists were thus inhibited from speculating about their shapes and sizes and the way in which these might be used to explain chemical phenomena. In any case, such concepts did not provide the kinds of explanation which chemists needed.

Consequently chemists found themselves in effect on the horns of a dilemma, or rather more than one dilemma. If they used the notions of the nature of elements and principles which developed naturally from the requirements of chemistry, they were departing from the traditional philosophical concepts which were felt to be academically respectable. If, on the other hand, they used the mechanical explanations which were the alternative and more up-to-date way to intellectual respectability, they were offending against the canon which required them not to make any assumptions that were not supported by the evidence of their senses through observation and experiment; and in any case the mechanical explanations that were offered by natural philosophers did not in fact solve their essentially chemical problems. They needed, both for the sake of making chemistry manageable and in order to achieve intellectual respectability by the standards of the Enlightenment, to arrange the facts of chemistry in some sort of order; but they could not fit them into any pattern which was conceived *a priori* without again offending against the accepted canon of methodology. They also needed to try to make chemistry quantitative, and to try to find mathematical laws from which chemical events could be predicted, still without making any *a priori* assumptions, but if possible by finding patterns in observed phenomena, measuring observed quantities, and so arriving at mathematical laws by induction. The ways in which they developed concepts of chemical combination and of the nature of elements and compounds which fulfilled the special needs of chemistry; attempted to find a natural order which was not preconceived, but into which chemical substances fell in virtue of their own nature; and attempted to make quantitative measurements which would lead to mathematical laws, so asserting the autonomy of chemistry as an independent discipline in its own right, will be explored in the chapters which follow.

Notes

1. See, for example, A. Thackray, *Atoms and powers*, Cambridge, Mass., 1970. See also A. M. Duncan, 'Particles and eighteenth-century concepts of chemical combination', *British Journal for the History of Science*, **21**, 1988, pp. 447–53.
2. Anon. [Senac], *Nouveau cours de chymie*, 2nd edn, Paris, 1737 (1723), pp. lxxviii–lxxix.
3. French trans. of Margraf, *Opuscules chymiques*, 2 vols, Paris, 1762, Vol. I, pp. ii–v.
4. Anon. [P.-J. Macquer], *Dictionnaire de chymie*, 2 vols, Paris, 1766, Vol. 1, pp. xxii–xxiii.
5. P.-J. Macquer, trans. J. Keir, *Dictionary of chemistry*, London, 1771, introduction.
6. M. J. A. N. de Caritat, Marquis de Condorcet, *Oeuvres*, Stuttgart, 1968 (repr. of 1847 Paris edn), Vol. II, p. 130.
7. H. Pemberton, *A course of chemistry, divided into twenty-four lectures*, ed. James Wilson, London, 1771, pp. 10–11.
8. T. Thomson, *The history of chemistry*, London, 1830-1, 2 vols, Vol. I, p. 264, p. 303, and p. 235.
9. Jan Golinski, *Science as public culture: chemistry and Enlightenment in Britain, 1760-1820*, Cambridge, 1992. However, Golinski's choice of Joseph Priestley as an example of a chemist who got a public reputation through natural philosophy is not altogether fortunate, as Priestley was chiefly known, indeed notorious, as a political pamphleteer and an unorthodox theologian, with an interest in artificial airs (i.e. gases) as a similarly peculiar subsidiary activity.
10. Cullen manuscripts in the Department of Special Collections, Glasgow University Library, Item 5, Box 1. Cullen's view of chemistry as an autonomous discipline is discussed by A. L. Donovan, *Philosophical chemistry in the Scottish Enlightenment*, Edinburgh, 1975, pp. 93-102, and is also referred to by Golinski, op. cit., p. 17.
11. On methods of induction, see in particular F. Bacon, *Novum organon*, London, 1620, and *New Atlantis*, London, 1624.

12. On primary and secondary qualities, see especially Galileo, *Opere*, 15 vols, Florence, 1842 ff, Vol. IV, pp. 333 ff. On Galileo's logic and methodology in general, see G. Galilei, *Tractatio de praecognitionibus et praecognitis and Tractatio de demonstratione*, ed. W. A. Wallace, Padua, 1988, with the bibliography at pp. lxxv-lxxix.
13. R. Descartes, *Discours de la méthode*, Leiden, 1637; *Meditationes de prima philosophia*, Paris, 1641.
14. Descartes's theory of matter is mostly in the *Principia philosophiae*, Amsterdam, 1644, Parts III and IV. See E. J. Aiton, *The vortex theory of planetary motions*, London, 1972, pp. 30 ff.
15. John Locke, *Essay concerning human understanding*, Oxford, 1690.
16. George Berkeley, *Treatise concerning the principles of human knowledge*, Dublin, 1710, Sections 9-15.
17. Etienne Bonnot de Condillac, *Oeuvres philosophieques*, ed. Georges Le Roy, Paris, 1947-51.
18. Thomas Hobbes, *Leviathan*, London, 1651, especially Chapters 1 and 2. On the controversy between Hobbes and Boyle see S. Shapin and S. Schaffer, *Leviathan and the air-pump: Hobbes, Boyle and the experimental life*, Princeton, 1985.
19. David Hartley, *Observations on man*, London, 1749, Vol. I.
20. Joseph Priestley, *Disquisitions relating to matter and spirit*, London, 1777, p. 25.
21. Particularly through Voltaire, *Les éléments de la philosophie de Newton*, paris, 1738. See Pierre Brunet, *L'introduction des théories de Newton en France au XVIIIe siècle*, Paris, 1931.
22. P.-H. T. d'Holbach, *Système de la nature*, Vol. I, Paris, 1770, esp. pp. 31 f.
23. See I. Kant, *Opus postumum*, in *Gesammelte Schriften*, 22 vols, Berlin, 1902–42, Vol. XXI, esp. p. 293.
24. Especially in R. Boyle, *The sceptical chymist*, London, 1661. On Boyle's style, see Shapin and Schaffer, op. cit., pp. 63–6.
25. See Owen Hannaway, 'Johann Conrad Barchusen (1666-1723)', *Ambix*, **14**, 1967, pp. 96-111.
26. Referred to by Ephraim Seehl, *A new improvement on the art of making the true volatile spirit of sulphur*, London, 1774, pp. 42-3; R. Dossie, *The elaboratory laid open, or, the secrets of modern chemistry and pharmacy revealed*, London, 1758, pp. 44–7, 158 f, 162. See J. R. Partington, *History of chemistry*, Vol. II, London, 1961, pp. 323–4 n.
27. Dossie, loc. cit.; J. Page, 'Receipts for preparing and compounding the principal medicines made use of by the late Mr Ward', London, 1763, p. 263 (in *The medical museum*, by Gentlemen of the Faculty, Vol. I, London, 1763, pp. 255–78).
28. The evidence for Roebuck and Garbett's venture is set out and discussed in A. and N. L. Clow, 'Vitriol in the Industrial Revolution', *Economic History Review*, **SV**, 1945, pp. 44–55, and *The chemical revolution*, London, 1952, Chapter VI, pp. 133–43.
29. The evidence from the Doldowlod letters for this episode is examined by Eric Robinson in A. E. Musson and E. Robinson, *Science and technology in the Industrial Revolution*, Manchester, 1969, Ch. X, pp. 352–71.
30. See A. and N. L. Clow, *The chemical revolution*, London, 1952, pp. 100–6.
31. The traditional picture of Leblanc as a victim of the Revolution is based on the memoir by his grandson, A. Anastasi, *Nicolas Leblanc, sa vie, ses travaux, et*

l'histoire de la soude artificielle, Paris, 1884, which in turn was based on the report by J. B. Dumas, 'Rapport relatif à la découverte de la soude artificielle', *Compte rendu des séances de l'Académie des Sciences*, **42**, 1856, pp. 553–78. Dumas was reporting on behalf of a commission which had investigated the claims of J. J. Dizé, who had undoubtedly worked with Leblanc, to a major share of the credit for the development of the Leblanc process. C. C. Gillispie, 'The discovery of the Leblanc process', *Isis*, **48**, 1957, pp. 152–70, has reassessed the traditional picture with the aid of some of the contemporary evidence, and also concluded that Leblanc's contribution to the development of the process was more or less accidental. However, my colleague Dr J. G. Smith, who has studied a fuller range of the evidence in *The heavy chemical industries in France*, Oxford, 1979, pp. 209 ff, has argued convincingly that Gillispie underestimates the level of Leblanc's understanding of chemistry. At any rate, there seems to be no doubt that the industrial process was guided by whatever theoretical understanding was available. See also W. A. Smeaton, 'Louis Bernard Guyton de Morveau, F.R.S., (1737–1816) and his relations with British scientists', *Notes and Records of the Royal Society of London*, **22**, 1967, p. 116, with a reference to the Banks correspondence (British Museum (Natural History)), Vol. III, pp. 61–4.

32. H. L. Duhamel du Monceau, 'Sur la base du sel marin', *Mémoires de l'Académie Royale des Sciences*, 1736, pp. 215–32.
33. A. F. de Fourcroy, *Annales de Chimie*, **9**, 1791, pp. 305–52. Discussed by W. A. Smeaton, *Fourcroy: chemist and revolutionary*, London, 1962, pp. 120–1.

2

Chemical affinity and attraction in the seventeenth and early eighteenth centuries

Early history of the concept of chemical affinity

The oldest explanation for chemical combination is probably that which attributes human emotions to the substances concerned. The pre-Socratic Greek philosopher Empedocles, for instance, attributed combination, and indeed all other natural processes, to Love and Strife.[1] That implied that chemical change was due to the combination of different substances that did not change their intrinsic nature on combining, rather than to change in the intrinsic properties of a single kind of matter, and also that it was substances which had some sort of kinship that combined. Empedocles, however, did not mean to imply that the universe had human emotions—his Love and Strife seem to have been abstract and impersonal forces.

In alchemical writing it was normal to speak of chemical reactions as if the substances involved were personified and moved by emotions of love and hate. In Chinese alchemy the central notion of the male and female principles, Yang and Yin, as the basis of the composition of all substances, being rooted in Taoism, is meant to be taken literally.[2] In Arabic alchemy Sulphur and Mercury, the active and passive principles, also represented the male and female principles, which are opposite in nature but attracted to each other, and whose perfect union produces gold, the complete metal towards which all other metals tend.[3]

In European alchemy, however, animistic language is usually allegorical, for alchemy was concerned with spiritual purification as much as, or even more than, with chemical transmutation. Indeed, it is often hard to know whether a passage of alchemical writing has a literal, chemical meaning at all as well as its allegorical meaning. The roots of the allegorical meanings lie far back in European thought, and are perhaps connected with Gnosticism and with the dualist heresies of the Manichaeans or the medieval Cathars. For the present purpose, however, we need not probe deeply into such origins, for they had little influence on natural philosophers of the Age of Reason.

Authors of the sixteenth and early seventeenth centuries who were not alchemists but were nevertheless still under the influence of the alchemical tradition from which they were breaking tended also to use words that implied that chemicals loved or hated each other. Presumably they did not really mean to imply that the substances in question were animated by human emotions, and their language is to be taken as metaphorical. yet they hardly seem to have been aware that it was metaphorical, and there is in such cases little indication of what the metaphor means. The distinction which a professional twentieth-century scientist would assume without question between the style appropriate to *belles lettres* and the austere jargon of scientific communication had not yet been thought of. The scientific profession did not yet exist, and natural philosophers did not write for professionals or even for specialists. They were writing for educated people in general who might wish to read about natural philosophy for its own sake.[4] Literary style of the period was given to elaborate and sometimes far-fetched metaphors, and writers on natural philosophy naturally tended in the same direction. Indeed, an air of mystery and a challenge to the reader to puzzle out the intended meaning or meanings would add to his enjoyment.

In early modern times, however, a slightly different explanation of chemical combination had appeared. It suggested that similar substances combined not because they had feelings for each other but because they were akin in composition. That is the chief meaning of the word 'affinity', which was originally a metaphor from family relationships between human beings, and was later criticized because it retained at least a flavour of animism.

The earliest use of the term in its chemical sense was probably in the *Liber mineralium* of Albertus Magnus (1193–1280). He wrote that sulphur blackened silver and 'on account of the affinity of its nature burns the metals'.[5] There the word seems to mean no more than similarity of constitution.

The author of the manuscript attributed to Geber, but probably written in the second half of the thirteenth century by a Central European chemist, explained the observation that there is a definite order of reactivity of metals by assuming similarity of composition.[6] Thus he is a forerunner of the eighteenth-century authors of tables of affinity. In the Arabic tradition, his theory is that metals consist of the philosophical principles Sulphur and Mercury in varying proportions. If the Sulphur and Mercury are pure and in the most perfect equilibrium, the resulting metal is gold. Any defects of purity or proportion produce silver, lead, tin, iron, or copper. It follows that purification or adjustment of the proportions of Sulphur and Mercury can convert any of these metals to gold. That was, of course, the aim of the alchemists in the superficial interpretation.

The proportion of Mercury in the metals decreases in the order gold, tin, silver, lead, copper, and iron, and this is also the order of their reactivity with mercury, according to pseudo-Geber. The order of their reactivity with sulphur and of their tendency to be calcined by fire is the reverse. 'That body is more burnt, which is less nigh to the Nature of the Perfect.'

The work of Paracelsus (*c.*1493–1541) cannot always be interpreted as if he were referring simply to the material world, as he believed that the occult world had powerful influences on the visible world and that it was the business of the physician to discover and direct those influences. However, he seems to have been speaking in terms of the ordinary visible world when he observed, like pseudo-Geber, that the order of reactivity of metals with mercury depended on the degree of similarity of their nature to that of mercury. 'Then the quicksilver has the nature and property that it amalgamates with the metals and completely and utterly unites with them, yet with one much more readily than with another, according to which has the closest affinity of nature to it.'[7] The order of reactivity according to Paracelsus is gold, silver, lead, tin, copper, and iron, followed by alloys according to the proportions of their components.

Seventeenth-century concepts

In the seventeenth century the notion, rather like that of dependence on resembling mercury which we have just encountered, that metals react more or less readily according to whether their nature is more or less metallic, is found in several authors. Jean Béguin (1550–1620) in the 1615 edition of his *Tyrocinium chymicum* explains the reaction by which butter of antimony (the deliquescent chloride of antimony) may be prepared from crude antimony sulphide and mercuric chloride as follows:

> I say that the vitriolic spirit has an extreme sympathy with the metals, and more or less with the other minerals according to how close or distant they are from the metallic nature. And because the regulus of Antimony [i.e. the metal] is closer to the metallic nature than the Mercury, that is why in distilling the Mercury sublimate with antimony, the vitriolic spirit of the Sublimate leaves the Mercury and joins and attaches itself to the regulus of Antimony.[8]

(To Béguin, as to others of his time, all mineral acids are 'vitriolic spirit'.)

The word by which Béguin in this passage denotes that tendency to combine which is explained by the more truly metallic nature of antimony is not *affinité* but *sympathie*, which is no doubt an echo of alchemical vocabulary and implies emotion, metaphorically perhaps rather than literally; whereas affinity is not necessarily an anthropomorphic concept, but implies merely some relationship, which may be similarity of constitution. Such animistic language is found also in the *Novi furni philosophici* of Johann Glauber (1604–70), where failure to combine is explained by hatred as well as combination by love, as in the following passage:

Nothing more prosecutes Gold with an hostile hatred than burning sulphur and sulphurious salts, such as are Alcalies, and crude tartar; the cause of this hatred is, because Gold is nothing else, but a fixed Sulphur, and therefore it disagreeth by a capital hatred, with every burning Sulphur, Silver and Lead do love every Sulphur, and all sulphureous Salts, such as are Vitriol, Salt Peter, Sal Armoniack, and the like, the which they stand in need of for their colour; and they have an hatred against Kitching Salt, because it is of a mercurial nature, and therefore not requiring its help, but only desiring a Sulphur and Tincture. Copper, Iron and Argent vive, or Quick-silver do possess both natures, to wit, a mercurial and sulphureous one, and for that cause they prosecute all Sulphurs and any Salts with love.[9]

Packe, the English translator of Glauber, uses the word 'affinity' in the following passage, which refers to cupellation:

[The Alchymist supposes] also that the trial of tin, copper and iron, made in a cuple with lead to be that true genuine bath thereof; not observing that lead hath no affinity with iron, and tin in a stronger fire, but to reject what is black, and unclean, without any perfection.[10]

Otto Tachenius (*fl.* 1666), who developed the observation of Franciscus Sylvius on the importance of reactions between acids and alkalis into a theory that everything in the universe was composed of acid and alkali, interpreted their union in terms of love and hatred. However, the following passage makes it clear that he consciously intended such phraseology to be taken metaphorically. After claiming that Hippocrates had reduced what he refers to as Aristotle's three causes to two necessary and sufficient principles, calling them Fire and Water, he continues:

Others stile them Lis and Concordia, Attraction and Repulsion, Rarefaction and Condensation, Male and Female, etc. But I for the clearer knowledge and explanation of them, do call those two Principles of Hippocrates, Acid and Alcaly, because all things in the Universe are made up of those two Universal Principles, as I shall gradually show by Experience ... Hence arose Sal, Sulphur, and Mercury, the Three principles of some philosophers.[11]

Naturally, philosophers who admired the mechanical philosophy and wanted to introduce it into chemistry objected to such emotions as love and hatred being attributed to inanimate substances. They disapproved of imposing systems conceived *a priori* to explain phenomena, and preferred simple, material explanations which they could claim to have inferred from the phenomena without preconceptions. Even in the first half of the seventeenth century J. B. van Helmont (1579–1644), for instance, who had not moved very far from the old iatrochemistry and alchemy towards the new notions, denied that chemical change was due to sympathy or antipathy between the substances involved, though the concept of fermentation which he suggested instead was not much less obscure. He wrote: 'Nevertheless no alteration or transmutation takes place in things through the longings of matter which are imagined, but only by means of ferment.'[12]

On the other hand, quite late in the seventeenth century there were still writers on chemistry who used metaphorical or animistic language. One such was John Mayow (1641–1679), though in some other ways he was much more in tune with the new mechanical philosophy than Glauber. His theory, which owed much to the work of Hooke and Lower, was that air consisted of two kinds of particle, the nitro-aerial particles which formed the pabulum of fire and lodged in solids as they burnt, and particles of another kind which did not take part in combustion. His concept of 'nitro-aerial particles' has been compared with the much later concept of oxygen, though there are fundamental differences and the attempt to identify the two concepts is an anachronism: the meaning and context are different. The heat involved in combustion was held to be due to the violent collision between the sulphureous particles, which all inflammable substances contained and which were actually responsible for combustion, and the nitro-aerial particles.

He writes in one passage, for instance, of the nitro-aerial spirit's having 'the saline particles ... firmly clasped in its bosom',[13] and of 'a great affinity and likeness among all acid salts'. Then he says that

> The particles of nitrous spirit generated in the earth in the manner aforesaid, as soon as they are produced, approach the seeds of the fixed salts, which, as has been elsewhere shown, are hidden in the bosom of the earth, and solicit and call them forth into conjugal union as a suitable consort and of their own kin.[14]

A little later, in discussing elements, he says

> Nitro-aerial spirit and sulphur are engaged in perpetual hostilities with each other, and indeed from their mutual struggle when they meet and from their diverse state when they succumb by turns all the changes of things seem to arise ... Salt has great affinity and relationship with nitro-aerial spirit and also with sulphur; for these very active elements are by turns married to salt as to a fitting bride, and are fixed in its embrace, as will presently be shown.[15]

Presumably Mayow did not mean this talk of brides and embraces and conjugal union to be taken literally; but his use of it enables him to avoid giving any literal explanation of what he does mean. All that is clear is that he supposes the reason for two substances' combining with each other in preference to others to be some sort of intrinsic affinity. Elsewhere, however, he speaks as if he were a sound mechanical philosopher. Always he seems to assume that matter is particulate, and that the properties of the substances depend on the shape, arrangement, and motion of their particles, in the manner of his contemporary Boyle, even recalling Descartes in his idea of the particles wearing each other down. He insists that 'the form of flame is chiefly due to the nitro-aerial spirit set in motion',[16] and continues:

Further it should be noted that the flame of kindled sulphur, as indeed flame of every kind, consists in this that the sulphureous particles of the deflagrating substance and the nitro-aerial particles mutually excite themselves to a very rapid motion, as we have shown above. But as the minutely divided saline particles of the sulphur are very closely united to its sulphureous particles, it happens in the deflagration of sulphur (when the sulphureous and nitro-aerial particles throw each other into fiery motion) that the saline particles of the sulphur, adhering to its sulphureous particles are by the frequent impacts of the nitro-aerial particles struck, rubbed and comminuted, so that the saline particles from being often rubbed and pounded, are at last sharpened like small swords and are moreover so attenuated as to be changed from rigid and solid into flexible and fluid particles.[17]

Mayow describes what later came to be known as elective affinity in somewhat figurative language, but in such a way as to make it clear that he has correctly observed the phenomenon even if he cannot explain it. He writes, for instance,

As acid salts leave volatile salts to form a closer union with the fixed salt or tartar [potassium carbonate] as being a more suitable partner, so doubtless fixed salts select some one acid in preference to others that they may combine with it in a closer union.[18]

Several other examples follow.

Boyle

In contrast, Robert Boyle (1627–91), whose controversy with Hobbes was mentioned in the previous chapter, wholeheartedly rejected animistic or metaphysical interpretations in favour of mechanical models in his explanations of the causes of chemical combination. Thus he wrote in his 'Reflections upon the hypothesis of Alcali and Acidum':

I look upon amity and enmity, as affections of intelligent beings, and I have not yet found it explained by any, how those appetites can be placed in bodies inanimate and devoid of knowledge, or so much as sense. And I elsewhere endeavour to show, that what is called sympathy and antipathy between such bodies does, in great part, depend upon the actings of our own intellect, which supposing in everybody an innate appetite to preserve itself both in a defensive and an offensive way, inclines us to conclude, that the body, which, though designlessly, destroys or impairs the state or texture of another body, has an enmity to it, though perhaps a slight mechanical change may make bodies, that seem extremely hostile, seem to agree very well and co-operate to the production of the same effects.[19]

This passage sums up very well the objections which those who accepted the mechanical philosophy felt to the implications of such words as 'affinity' if they were to be taken literally. Boyle, indeed, explained even magnetic and electric attraction mechanically. He supposed that the magnet or the electrically charged body must emit an effluvium which pushed the other body towards or away from it.[20] Like Descartes, he objected to the suggestion of action at a distance. It seemed evident to him that one body could not move another unless they were in contact. In 'An enquiry into the cause of attraction by suction' he argued that attraction must be merely a species of pulsion, in which the mover was in direct contact with the thing moved, and pushed it.[21] He accepts that there is some empirical basis for the idea that like combines with like, or as it was sometimes expressed, 'like rejoices in like' (*simile simili gaudet*), but does not accept that it explains their combination.

Most of the chemists pretend, that the solutions of bodies are performed by a certain cognition and sympathy between the menstruum [solvent] and the body it is to work upon. And it is not to be denied, that, in diverse instances, there is, as it were, a consanguinity between the menstruum and the body to be dissolved; as when sulphur is dissolved by oils, whether expressed or distilled: but yet, as the opinion is generally proposed, I cannot acquiesce in it.[22]

On the contrary, he subscribes to a mechanical explanation.

As for what is commonly said, that oils dissolve sulphur, and saline menstruums metals, because (as they speak) *simile simili gaudet*: I answer, that, where there is any such similitude, it may very probably be ascribed, not so much with the chemists, that favour *Aristotle*, to the essential forms of the bodies, that are to work on each other; nor, with the mere chemists, to their salt, or sulphur, or mercury, as such; but to the congruity between the pores and figures of the menstruum, and the body dissolved in it, and to some other mechanical affections of them.[23]

Boyle does try to explain the selectiveness of chemical affinity (a task which few others attempted) by the shapes of the particles concerned. In 'Reflections upon the hypothesis of Alcali and Acidum', for example, he wrote:

Mercury has corpuscles of such a size and shape, as fit them to insinuate themselves into the commensurate pores they meet with in gold, but make them unfit to enter readily the pores of iron, to which nature has not made them congruous; as on the other hand the saline corpuscles of aqua fortis [concentrated nitric acid] will easily find admission into the pores of iron, but not into those of gold, to which they do not correspond as they do to others.[24]

The *Cours de chymie* of Nicolas Lemery (1645–1715), which was published in the same year (1675), gives rather similar explanations in terms of the shapes of the particles of matter, based no doubt on the ideas of Descartes, though their origin goes back to the Greek atomists.[25]

In his 'Suspicions about some hidden qualities of the air', Boyle shows some understanding of the importance of relative quantities as well as affinities in determining the course of a reaction. Attention was again directed to this point by Bergman a century later, and it proved to be of great importance to Berthollet's iconoclastic theories and of course in nineteenth- and twentieth-century chemistry.

The reaction discussed by Boyle in this instance is evaporation, which he sees as a kind of solution of the evaporating substance in the atmosphere.

> In divers cases, the quantity of a menstruum [solvent] may much more considerably supply its want of strength, than chemists are commonly aware of ... And as to those bodies on which the aerial menstruum can though but slowly work, the greatest quantity of it may bring this advantage, that whereas even the strongest menstruums, if they bear no great proportion in bulk to the bodies they are to work on, are easily glutted and being unable to take up any more, are fain to leave the rest of the body undissolved, our aerial menstruum bears so vast a proportion to the bodies exposed to it, that when one portion of it has impregnated itself as much as it is able, there may still come fresh to work further on the remaining part of the body.[26]

It will be noticed that even Boyle has used a metaphor in the word 'glutted', though no doubt he was unaware of it, and that this conceals the absence of any mechanical explanation of such 'glutting'.

Newton

In the present work we are concerned with eighteenth-century Newtonianism, that is with the influence of what the eighteenth century believed to be the ideas of Isaac Newton (1642–1727), rather than with Newton's private thoughts which lay behind his published writing. He was a very private man, and recent research has begun to uncover some of his unpublished speculations about the ultimate causes and hidden mechanisms of the workings of nature. Certainly he believed that God was omnipresent and responsible for all physical events: the philosopher's task was to understand the mechanisms by which God chose to carry out His will. One of Newton's aims was evidently to recover the *prisca theologia*, the lost knowledge which the ancient world had of the hidden powers of Nature. Although he was as thoroughly convinced as was Boyle that matter as we know it is particulate, he associated with its particles powers of attraction and repulsion which according to the strict mechanical philosophy could well be counted as occult powers. In Newton's mind they seem to have been among the concealed active principles of Nature which were linked with his thoughts on metaphysical philosophy and theology as well as with his understanding of alchemy. In the Leibniz–Clarke debate, and to such eighteenth-century thinkers as Hutchinson or James Hutton, these deeper implications of Newton's ideas were certainly important.

Thus alchemy, and its connections with the mystical and Hermetic strains in Renaissance thought, were a potent influence on Newton's thoughts about the hidden mechanisms of Nature and the fine structures of matter. Very large quantities of notes survive in which he has transcribed, translated, indexed, or summarized a great bulk of alchemical writing, and perhaps added a very few occasional contributions of his own. It would, of course, be an anachronism to expect him to have found alchemy absurd or unscientific. Criteria of what is absurd and what is common sense are not absolute, but depend on the fashions of the time; and there was nothing in seventeenth-century mechanical interpretations of the nature of matter to suggest that gold was not a compound which might be synthesized. Probably Newton was trying to discover the practical interpretation of the elaborately metaphorical language of alchemy. At any rate, the alchemical texts must have been a rich source of material for Newton's study of chemistry.[27]

Nevertheless, although Newton's theories could be interpreted in several different ways by eighteenth-century chemists, their knowledge of his ideas depended on what was available to them in published form. To them he was essentially a materialist, mechanical philosopher and mathematician. In what follows, Newton's work will be described mainly as it appeared to a chemist of the mid-eighteenth century, and the concealed depths of his thinking as revealed in his surviving manuscripts will not be much explored. Many enlightened eighteenth-century, and indeed later, natural philosophers or scientists would have been horrified by them.

Certainly Newton's interest in chemistry lasted the whole of his adult life.[28] The paper 'De natura acidorum', which contains a clear account of his theory of chemical mechanisms, was stated by John Harris, in whose *Lexicon Technicum* it was published, to have been written in 1692; but the supplement to the *Lexicon Technicum* which contains it did not appear until 1710. Even then, Harris says that it was not published on Newton's instructions, but merely 'by his leave'.[29] It was only in the 31st Query (in the eventual numbering) at the end of the *Opticks* that Newton deliberately had printed an account of his views on chemical attraction, and that Query was first published in Latin in the 1706 edition of the *Opticks* and in English only in the 1717 edition. The ideas in the Query had a very great influence on eighteenth-century chemists; but they are expressed as a series of questions or suggestions, not in the form of a statement.

Also, the Query carefully avoids any suggestion about the mechanism of the causes of attraction and repulsion, as does the *Principia*. Yet, as we shall see, Newton did entertain various speculative theories on the subject, to which he occasionally referred in writing. In those theories he often made use of the concept of an ether, which is in fact employed in his speculations about the causes of such optical phenomena as refraction, conveyed in Queries at the end of the *Opticks* which are earlier than the 31st in the eventual numbering but which first appeared in editions later than the first appearance of the 31st. There has been much discussion of the reason for the curious coyness which Newton showed about publishing his theories. It is well known that he had a short period of mental illness in 1693, when he showed signs of persecution mania and accused John Locke, for instance, of spreading falsehoods about him. However, he recovered from that, though he was suspicious of plagiarism and sometimes resentful of criticism.[30] A more likely explanation of his attitude is less dramatic.

When in 1672 he was first persuaded to publish his work on colours he was exposed to a long controversy and to considerable criticism. Evidently he hated that, and dreaded criticism ever after. Yet he still liked to have his ideas known and praised. Halley did in the end manage to persuade him to publish his *Principia*. It was probably this ambivalence about publication which made him so careful about committing himself. He does not, however, seem to have had any doubt in his own mind whether his views were correct or not, or what his own views were at any particular time.

The main theories of the *Principia* emerge from an analytical treatment of mathematical principles. Newton could fairly claim for that work '*hypotheses non fingo*' ('I construct no hypotheses'), as he says specifically about possible mechanical explanations of gravity. The application of the theories to physics and astronomy was soundly based on well-accepted experiments and observations. Thus he guarded himself against the kind of criticism which was considered most damaging in the Royal Society at the time, that a philosopher was speculating about systems and theories which were not soundly based on experiment. He avoided the possibility of any such charge against his concept of gravitation, and also against any suggestion that he was thinking of anything like an Aristotelian form or quality, by the following disclaimer:

> I here use the word attraction in general for any endeavour whatever, made by bodies to approach to each other, whether that endeavour arise from the action of the bodies themselves, as tending to each other or agitating each other by spirits emitted; or whether it arises from the action of the ether or of the air, or of any medium whatever, whether corporeal or incorporeal, in any manner impelling bodies placed therein towards each other.[31]

The Queries at the end of the *Opticks* eventually numbered 30 and 31 were both first published in the Latin edition of 1706, and those eventually numbered 17 to 22 not until the second English edition of 1717. Dr Thackray has suggested that in the 1706 edition Newton had been driven to defend his philosophy of matter against the attacks of Leibniz by 'setting out more fully the role of attractive forces in the economy of Nature'.[32] By that time, also, Newton's position at the summit of the scientific community in England was unassailable, and he no longer needed to be so cautious about risking criticism. Many of his earlier critics were dead—for instance Robert Hooke had died in 1703. Perhaps both of these were among the reasons for which Newton felt willing to publish the suggestions in the Queries, and to allow the paper 'De natura acidorum' to appear in 1710.

Although his suggestions in the 31st Query are put in the form of questions, his followers in the eighteenth century supposed that he thought the suggestions at least highly probable, and they were clearly right. Perhaps those questions which are in the indicative, starting with the words 'Are not ...?' or 'Have not ...?' are those about which Newton was more nearly certain than those which are in the subjunctive, beginning with the words 'May not ...?'

The Query begins with a paragraph which provided a model for most chemists in the mid-eighteenth century, and which recalls words previously used in the *Principia*.[33]

> Have not the small particles of Bodies certain Powers, Virtues or Forces, by which they act at a distance, not only upon the Rays of Light for reflecting, refracting and inflecting them, but also upon one another for producing a great Part of the Phaenomena of Nature? For it's well known, that Bodies act upon one another by the Attractions of Gravity, Magnetism and Electricity; and these Instances show the Tenor and Course of Nature, and make it not improbable but that there may be more attractive Powers than these. For Nature is very consonant and conformable to her self. How these Attractions may be perform'd, I do not here consider. What I call Attraction may be perform'd by impulse, or by some other means unknown to me. I use that Word here to signify only in general any Force by which Bodies tend towards one another, whatsoever be the Cause. For we must learn from the Phaenomena of Nature what Bodies attract one another, and what are the Laws and Properties of the Attraction before we enquire the Cause by which the Attraction is perform'd. The Attractions of Gravity, Magnetism and Electricity, reach to very sensible distances, and so have been observed by vulgar Eyes, and there may be others which reach to so small distances, even without being excited by Friction.[34]

In the first sentence of this paragraph Newton appears to be suggesting that the particles have 'Powers, Virtues or Forces by which they act at a distance' as intrinsic properties. However, the later sentences make it clear that he is not here suggesting either that these powers are intrinsic or that they are caused by some extrinsic mechanism. He is concerned merely with explaining phenomena which can be observed experimentally, and he disclaims any hypotheses about the causes of attraction.

After this opening paragraph he suggests explanations of a number of chemical reactions through this theory of attraction. He attributes the deliquescence of salts of tartar (potassium carbonate) to an attraction between the particles of the salt and the particles of water vapour floating in the air, and the fact that some other salts are not deliquescent to their lack of such an attraction. However, salt of tartar does not draw water from the air in more than a certain proportion to its quantity 'for want of an attractive Force after it is satiated with Water'.[35] Thus Newton has picked out two phenomena—the selectiveness of chemical attraction and saturation—which any theory of the mechanism of chemical combination would have to explain if it were to succeed, but for which in fact his theory offers little explanation. The neutralization of electrostatic charges might have provided a parallel for neutralization in chemistry, or in eighteenth-century terms the saturation of chemical attraction, though not to its selectiveness; but Newton does not refer to that possibility.

Next he argues that the evolution of heat during chemical action is evidence of very violent motion due to powerful forces of attraction. Boyle had also emphasized the motion of his corpuscles, and Newton's language in this passage recalls that of his letter to Boyle of 1679 (quoted below) where he wrote (of the pressure of the aether) 'then will that excess of pressure drive them with violence together'.[36] There are several other phrases in this Query which recall the Letter.

Newton goes on to consider several cases where a compound is dissociated because one of the substances in it is more strongly attracted by some other substance than by the one with which it is originally combined. He points out that *Lapis Calaminaris* (zinc carbonate) precipitates iron from a solution in *aqua fortis* (concentrated nitric acid), iron precipitates copper, copper precipitates silver, and iron, copper, tin, or lead precipitates mercury; and he suggests that these reactions are also the result of differences in the strength of the attraction between the acid particles and the particles of the various metals.[37] This order of reactivity is similar to those given by pseudo-Geber and Paracelsus; but those authors are less likely to have influenced the eighteenth-century compilers of affinity tables than Newton. However, no explanation is given of the presumed differences in the strength of attraction between various kinds of particle.

Newton shows that 'Salts are dry Earth and watery Acid united by attraction',[38] in which he is following the general lines of Tachenius' theory of the nature of salts; and he is evidently aware that the properties of a compound are not simply a blend of those of its components, as Boyle had made clear. As an explanation of the fact that *aqua fortis* dissolves silver and not gold, whereas *aqua regia* (a mixture of concentrated nitric and hydrochloric acids) dissolves gold and not silver, a favourite topic for the authors of mechanical explanations, he suggests that '*Aqua fortis* is subtle enough to penetrate Gold as well as silver, but wants the attractive force to give it Entrance; and that *Aqua regia* is subtil enough to penetrate Silver as well as Gold, but wants the attractive Force to give it Entrance.'[39] This is one of the few places where Newton speculates in print about the actual size of particles and its relationship with chemical properties, as Boyle had often done.

He asks whether the diffusion of the particles of a salt or vitriol (sulphate) through water, although they are heavier than water, does not imply that they repel each other, or attract water more strongly than they do one another. From the fact that a salt on crystallization from solution forms regular shapes, he argues that 'the Particles of the Salt, before they are concreted, floated in the Liquor at equal distances in rank and file, and by consequence that they are acted upon by some Power which at equal distances is equal, at unequal distances unequal.[40] Also, from the fact that the particles of Iceland spa 'act all the same way upon the Rays of Light for causing the unusual Refraction', he asks, 'may it not be supposed that in the Formation of this Crystal, the Particles not only ranged themselves in rank and file for concreting in regular Figures, but also by some kind of polar Virtue turned their homogeneal sides the same way?'[41] This line of thought is developed in the suggestions of Query 26 about polarized light.[42]

Newton next considers cohesion. As an explanation of it he rejects the idea of 'hooked Atoms, which is begging the Question'.[43] By that he may mean that ideas of the individual properties of unobservable particles ought to be derived from knowledge of the nature of cohesion, rather than the other way round. If so, the implied criticism of Boyle is a little unfair, for Newton has just been speculating about the sizes of such particles himself, while discussing the solution of gold and silver in acids. It is more likely that he means that such an explanation fails to account for the particles' coming together in the first place, and explains only how they are held together once they are in contact. His theory is dynamic, Boyle's is static.

Newton also rejects the Cartesian doctrine that 'Bodies are glued together by rest, that is, by an occult quality, or rather by nothing', or that 'they stick together by conspiring motions, that is, by relative rest among themselves', and prefers to infer from the cohesion of bodies that 'their Particles attract one another by some Force, which in immediate Contact is exceeding strong, at small distances performs the chymical Operations above mentioned, and reaches not far from the Particles with any sensible Effect'.[44] The implied judgment that the assumption of such otherwise unobservable forces is more rational, less speculative and subjective, and self-evidently more acceptable than the Cartesian explanations which he rejects, is plainly a subjective judgement, as indeed such judgements must always be in the end.

Since even liquids and vapours can become solid, the particles of all substances must be hard; and since hard bodies are very porous, and are composed of particles which can touch at only a few points, the simple particles themselves must be much harder. Newton adds that it is difficult to conceive how such particles can stick together so firmly 'without the assistance of something which causes them to be attracted or pressed towards one another'—one of the few hints in the Query that the cause of attraction may be external.[45]

Various cases of capillary rise are then considered and attributed to attraction between particles. Newton is able to estimate the force of this attraction quantitatively. He mentions some experiments of Hauksbee's showing that attraction of this sort is very great when the distance is very small and decreases in inverse proportion to the distance. Here again there is a hint that the cause of such an attraction may be external: 'There are therefore Agents in Nature able to make the Particles of Bodies stick together by very strong Attractions. And it is the Business of experimental Philosophy to find them out.'[46]

Thus the smallest particles may cohere by the strongest attractions, forming bigger particles with smaller attractive power; and so on until the progression ends in the biggest particles 'on which the Operations in Chymistry, and the Colours of natural Bodies depend, and which by cohering compose Bodies of a sensible Magnitude'.[47] The physical properties of substances depend on whether their parts slide upon each other or bend and yield without sliding, and so forth. Again Newton's suggestions in this passage recall Boyle's theory of the nature of matter. The concept of a hierarchy of compound particles was taken up by several other chemists.

Newton next considers in detail the idea of repulsion between particles. 'As in Algebra, where Affirmative Quantities vanish and cease, there negative ones begin; so in Mechanics, where Attraction ceases, there a repulsive Virtue ought to succeed.'[48] The existence of such a repulsive force, he argues, is supported by the phenomena of light, since he explains refraction and reflection as due to the repulsion of the refracting or reflecting substance for the particles of rays of light. Similarly repulsion is suggested by the fact that vapours and gases are evolved from solid bodies by heat or fermentation, and greatly expand. Mechanical explanations of this expansion 'by feigning the particles of Air to be springy or ramous, or rolled up like Hoops, or by any other means than a repulsive Power'[49] (Boyle's explanation, in fact, though Newton does not say so), are rejected as unintelligible. Newton's analogy between the continuity of the series of positive numbers and negative numbers and the continuity of attraction and repulsion could not be taken very far, since it is only in an arithmetic series that the terms gradually dwindle to zero and then become increasingly larger negative numbers, and attraction varies as a power series. However, the suggestion turned out to be fruitful in the long run, for example in the work of Boscovich.

The Query ends with some general considerations about natural processes and scientific methods, in which Newton defends his theory of attraction against the natural charge that it was merely speculation about unobservable 'occult qualities'. It is quite another thing, he argues, to 'derive two or three general Principles of Motion from Phaenomena, and afterwards to tell us how the Properties and Actions of all corporeal Things follow from those manifest Principles', which 'would be a very great step in Philosophy, though the Causes of those Principles were not yet discover'd'.[50] Whether Newton's own use of the notion of attraction could be said to constitute the derivation of 'two or three general Principles of Motion from Phaenomena' is a matter of opinion. Presumably the results rather than the logical basis may be said to justify it. Nevertheless, the grandeur with which he tried to bring the whole workings of the universe within his few general principles has inspired readers of the *Opticks* ever since, and enables us to understand the hold which such ideas had over his imagination. The principles are the principles of inertia, the cause of gravity, the cause of fermentation, and the active principles of the particles of matter. At any rate, to eighteenth-century followers of Newton it seemed plain that his style of reasoning was justified. Each age has its own conventions of rationality.

In the *De natura acidorum*, which is mentioned much less often than the 31st Query by later chemists, the notion that the particles of a substance have an attractive force is applied in a rather similar way to explain solution and chemical reactions. No explanation is offered why a particular kind of particle should be attracted by some particles but not by others, or in other words for the selectiveness of chemical attraction. However, an explanation of saturation is briefly sketched:

> If these Acid Particles be joyn'd with Earthy ones, in but a small Quantity, they are so closely retain'd by them, as to be quite suppressed and hidden as it were by them; so that they neither stimulate the Organ of Sense, nor attract Water, but compose Bodies which are not Acid, *i.e.* Fat and Fusible Bodies, such as are *Mercurius dulcis* [mercurous chloride], *Common Brimstone* [sulphur], *Luna Cornea* [fused silver chloride], and *Copper* corroded by *Mercury Sublimate* [mercuric Chloride].[51]

There is also another reference to the notion of a hierarchy of particles of increasing complexity:

> All bodies have Particles which do mutually attract one another: The Summs of the least of which may be called Particles of the *first Composition*, and the

Collections or Aggregates arising from them primary Summs; or the Summs of these Summs may be called Particles of the second Composition, etc.[52]

It is easy to believe that the *De natura acidorum*, and consequently the ideas underlying the 31st Query of the *Opticks*, date back to 1692, as Harris says; for the general view of the nature of matter on which they are based was expressed by Newton at various times throughout his life. Similar ideas occur in the suppressed *Conclusio* written for the *Principia*, probably in 1687.[53] The main result as far as chemistry was concerned was that in addition to the still vaguely animistic concept of affinity, and the purely mechanical interpretation of chemical combination in terms of the shapes and sizes of particles, a third kind of explanation was available. Although, as we shall see, affinity and attraction came gradually to mean the same thing as defined by chemists, at first the notion that substances combined because of an observable force of attraction between their particles, and the kind of thinking about reaction mechanisms which Newton originated, seemed, in communities where Newtonianism was accepted, both more rational than the concept of affinity and less speculative than the older mechanistic views. The notion of chemical repulsion was also taken up by some later writers on chemistry, but much less often than that of attraction.

Newton's views on the cause of chemical attraction are more elusive. Nevertheless, it is appropriate to mention them here because they too had some influence on later thought when they became known. He was, of course, at pains to defend himself against the charge made by Cartesians and by Leibniz that attraction was an 'occult power', or in other words that it did not clearly relate the observed phenomena to the primary properties of matter.[54] Much of the defence against Leibniz was done by Clarke in the famous correspondence between the two; but there is little doubt that Clarke's advocacy had Newton's approval. At times Newton did seem to mean that attraction was an intrinsic property of the particles of matter. Nevertheless, he expressly disavowed that view, presumably because it would have seemed to justify the charge that attraction was an 'occult power'.[55]

If attraction was not due to some power residing in the particles of matter themselves, the alternative was naturally to attribute it to something outside the particles. The normal view of mechanical philosophers, when an object seemed to exert a force on another object not in contact with it, was that the apparently empty space between them must contain some ether or effluvium by which or through which the force was transmitted. Apart from Descartes' ether, Newton must have known Boyle's theories about effluvia.

In his manuscript *De aere et aethere*, written probably between 1673 and 1675, Newton attributes attraction and repulsion to the action of particles of the atmosphere,[56] though later he entertains similar hypotheses about the particles of the aether. The transfer is not as surprising as it may seem. The Latin word *aether*, like the Greek αἰθήρ from which it is derived, means 'air', in particular the pure upper air, as Newton must have known; and in Jakob Bernoulli's dissertation of 1683 the word means 'air' as opposed to 'aether'.'[57] However in the letter to Oldenburg of 7 December 1675, Newton has turned to the Hypothesis of 'an aethereall Medium much of the same constitution with air, but far rarer, subtiler and more strongly elastic'.[58] Air was, of course, normally assumed to be an element at this period. As evidence of such an aether he describes the experiment in which a glass electrified by rubbing would attract pieces of paper, which rather like Gilbert he attributes to the movement of an 'aetherial wind' expelled from the glass by the rubbing and condensing into it again. By analogy with this he supposed that the gravitational attraction of the earth might be due to the downward pressure of

> some other such like aethereall Spirit, not of the main body of flegmatic aether, but of something very thinly and subtily diffused through it, perhaps of an unctuous or Gummy, tenacious and Springy nature, and bearing much the same relation to aether, which the vital aethereall Spirit requisite for the conservation of flame and vitall motions (I mean not the imaginary volatile saltpeter) does to Air.[59]

When it reaches the earth, this aethereal substance condenses and is absorbed, so that fresh supplies of it continually stream down to take its place; and at the same time part of what is absorbed is again exhaled and rises again to the upper atmosphere. During its descent it would bear down with it the bodies which it pervades with a force proportional to the 'superficies of their parts'.[60] This theory is evidently much influenced by the Cartesians.

A little later Newton used the concept of an aether to explain attraction in a different way, in his letter to Boyle dated 28 February 1678/9. However, the letter was not published until 1744, so that it had much less influence on eighteenth-century thought than the suggestions published in Newton's lifetime. In the letter Newton suggested that 'there is diffused through all places an aethereal substance, capable of contraction and dilatation, strongly elastic, and, in a word, much like air in all respects, but far more subtle'. He continued 'I suppose this aether pervades all gross bodies, but yet so as to stand rarer in their pores than in free spaces, and so much the rarer, as their pores are less.'[61] He then went on to show how this supposition could be used to explain a variety of observations, including refraction, capillary rise, cohesion, and the action of solvents.

When two bodies come close to each other, the aether in the space between them 'cannot move and play up and down so freely in the strait passage between the bodies', and so will be rarer at a given distance from the surface of each body than when they are not close together. Consequently the aether will resist the bodies' approach to each other and will tend to force them further apart. However, when the two bodies are brought so close together that the excess of pressure of the external aether surrounding them over that of the rarefied aether between them is great enough to overcome 'the reluctance, which the bodies have from being brought together', they will rush together violently, and adhere strongly to one another.[62] Hence the hypothesis explains both the existence of a strong attractive force between bodies separated by less than a certain minimum distance, and a strong repulsive force between bodies at distances greater than this minimum.

Newton then proceeded to apply the hypothesis to explain various observed facts about solution. He referred to the fact that some solvents will dissolve certain substances but not others. 'So aqua fortis [concentrated nitric acid] dissolves silver, not gold, aqua regia [a mixture of nitric and hydrochloric acids] gold not silver, etc.' However, as elsewhere, he could not explain these differences, and was reduced to saying, 'There is a certain secret principle in nature, by which liquids are sociable to some things and unsociable to others.'[63] This sociability, though it may be meant as a metaphor, is exactly the kind of 'affection of intelligent beings' which Boyle had objected to, and it is quite inconsistent with the mechanical philosophy. It recalls rather the language of alchemy, which was so familiar to Newton. Underlying his acceptance of his own variety of mechanical philosophy, in fact, Newton had a profound belief that there were a small number of secret principles in nature which were the means by which observable phenomena were brought about. Clearly he cherished a hope that these hidden principles might be revealed, though he took great care not to speculate about them.

These letters would no doubt be made public at meetings of the Royal Society; but Newton always implies in them that he is putting forward interesting and possible hypotheses for the sake of discussion rather than stating what he believes to be definitely so.

In 1687 Fatio de Duillier, a young Swiss who had spent some time in Holland and had been in touch with Huygens there, came over to London; and in 1688 he was elected to the Royal Society and gave an account of Huygens's then unpublished theory of gravitation. However he soon came under the influence of Newton's theories. Early in 1690 he expounded a theory of his own, in which he tried to explain the inverse square law of gravitation by the impact of the particles of a very finely dispersed aether on the particles of matter. Fatio claimed that Newton accepted this theory, although the only independent evidence of that acceptance seems to be in a rejected draft addition to the *Principia*.[64] The theory was, indeed, highly speculative, and it seems unlikely that

Newton would have committed himself in print to accepting it as more than a plausible hypothesis.

From about 1690 until after the publication of the Latin edition of the *Opticks* in 1706 there is no record of further speculations by Newton about aethers. Professor Guerlac has shown that Newton's return to such speculations in the *Scholium generale*, in the second edition of the *Principia* of 1713, and in the new Queries in the second English edition of the *Opticks* in 1717, was associated with Francis Hauksbee's experiments on electrical attraction and electro-luminescence, which Newton associated with an aether.[65] However, Miss Hawes has made the additional suggestion that although Hauksbee's experiments must have provided an added stimulus to Newton's speculations, the latter's main reason for returning to an aether theory was that such a theory could guard him from criticisms from followers of Leibniz or Descartes that his doctrine of gravitation involved action at a distance.[66]

The *Scholium generale* which was added to the second edition of the *Principia*, of which earlier drafts exist, begins with a paragraph on the difficulties in the way of Descartes's theory of vortices. Newton thus emphasizes that he is not adopting precisely the Cartesian form of the aether theory. The paragraph before the last of the *Scholium* sums up his views on induction:

But hitherto I have not been able to discover the cause of those properties of gravity from phenomena, and I frame no hypotheses; for whatever is not deduced from the phenomena is to be called an hypothesis; and hypotheses, whether metaphysical or physical, whether of occult qualities or mechanical, have no place in experimental philosophy. In this philosophy particular propositions are inferred from the phenomena, and afterwards rendered general by induction. Thus it was that the impenetrability, the mobility, and the impulsive force of bodies, and the laws of motion and gravitation, were discovered. And to us it was enough that gravity does really exist, and act according to the laws which we have explained, and abundantly serves to account for all the motions of the celestial bodies, and of our sea.[67]

This cautious empirical attitude to the construction of theories was normal among eighteenth-century chemists. Finally, Newton goes on to refer in a very general way to the aether.

And now we might add something concerning a certain most subtle spirit which pervades and lies hid in all gross bodies; by the force and action of which spirit the particles of bodies attract one another at near distances, and cohere, if contiguous; and electric bodies operate to greater distances, as well repelling as attracting the neighbouring corpuscles; and light is emitted, reflected, refracted, inflected and heats bodies; and all sensation is excited, and the members of animal bodies move at the command of the will, namely, by the vibrations of this spirit, mutually propagated along the solid filaments of the nerves, from the outward organs of sense to the brain, and from the brain into the muscles. But these are things that cannot be explained in a few words, nor are we furnished with that sufficiency of experiments which is required to an accurate determination and demonstration of the laws by which this electric and elastic spirit operates.[68]

This Scholium ranges very widely and leaves vague the mechanism by which the force and action of the subtle spirit cause the particles of a body to attract each other. The new Queries numbered 17 to 24 in the second English edition of the *Opticks* are a little more specific, though still no more than a sketch of a theory; and naturally they are chiefly about optics rather than chemistry. In Query 21 it is suggested that the aether is much rarer within dense bodies than in the space between them, and that its density everywhere increases with distance from solid bodies.[69] That would account for gravitation in a way which has something in common with Descartes's explanation of it, for bodies would naturally be impelled from regions of high density, that is of high pressure, to regions of low density and low pressure. That is very much like the letter to Boyle of 1679. Indeed, it is also slightly reminiscent of Descartes's mechanical explanation of gravity, that objects are impelled from the outer, faster-moving parts of vortices where the pressure is higher to the inner, slower-moving parts where the pressure is lower. Also,

If any one should suppose that *Aether* (like our Air) may contain Particles which endeavour to recede from one another (for I do not know what this *Aether* is) and that its Particles are exceedingly smaller than those of Air, or even than those of Light: The exceeding smallness of its Particles may contribute to the greatness of the force by which those Particles may recede from one another, and thereby make that Medium exceedingly more rare and elastick than Air, and by consequence exceedingly less able to resist the motion of Projectiles, and exceedingly more able to press upon gross Bodies, by endeavouring to expand itself.[70]

Mechanisms somewhat along those lines were applied to imponderable fluids such as heat and indeed phlogiston during the eighteenth century. However, the letter to Boyle written in 1679 was, as we have seen, unpublished until 1744; and the other more detailed manuscript ether theories that have been discussed above were not published until the twentieth century and were presumably unknown to chemists of the eighteenth century.

Thus eighteenth-century natural philosophers could adopt any one of several different approaches to the explanation of chemical attraction and still claim to be loyal followers of Newton; and, as we shall see in the next chapter, that is what most of them did. As far as the history of chemistry is concerned, that is all that need be said. Newton's private thoughts about alchemy and theology were forgotten, and would no doubt have been disavowed with horror as unsuitable for an age of enlightenment if anyone had remembered them. It was his published views which influenced chemists. In those, even though he was at times at least inclined to some kind of aether theory, he disclaimed any intention to speculate about the cause of chemical attraction, just as he disclaimed any notion that the powers of attraction or repulsion were intrinsic properties of the particles of matter and any theory involving action at a distance. He made it quite clear that the important parts of his work, about which he was certain, stood on their own without any theory of the cause of attraction. About such things as could not be established by experiment, he was not obliged to publish any theory at all.

Stahl

The audiences addressed by Newton and by Georg Ernst Stahl (1660–1734) were somewhat different. Newton's published works were intended both for the sophisticated audience of the members of the Royal Society and also for the rather wider audience of the educated upper and upper middle classes of England, and indeed to some extent of North-Western Europe. Admittedly very few could understand the mathematics of the *Principia*; but the *Opticks*, which after all was in English, was more widely understood. Stahl was court physician to the Duke of Saxe-Weimar, Professor of Medicine at the Pietist University of Halle, and then for the last eighteen years of his life court physician in Berlin. He wrote a much clumsier Latin than Newton, and for a rather less sophisticated audience, including medical men (though he himself denied that chemistry had any bearing on medicine), and an audience more likely to be occupied with practical chemistry, although his theories later enjoyed a wider popularity in Western Europe.

In comparison with the published work of Newton, much of Stahl's seems curiously old-fashioned. Most of his influence was through the general adoption of the phlogiston theory; but his ideas on affinity also influenced French chemists and some others. Some of his most important writings were translated into French by Baron d'Holbach (1723–89), a wealthy resident of Paris, an extreme materialist, and a contributor to the *Encyclopédie*. The theories of matter set out in Stahl's books seem sometimes inconsistent. However, that is chiefly because he deals with different levels of explanation at different times. In biology he was a vitalist, and in chemistry also he attacked some kinds of mechanical explanation, for instance in the preface to his *Fundamenta chymiae*.[71] However, what he meant to object to there was probably attempts to apply the concept of attraction, which was appropriate for physics, to chemical phenomena, and the implication that attraction was an inherent property of matter. Elsewhere he himself put forward explanations of chemical combination which may reasonably be called mechanical, and which do seem to describe features of the imperceptibly small particles of matter which are not directly accessible to experimental investigation.[72] That is curious, for Stahl's great virtue as a chemist was that he tried to put a theory of matter which was largely taken from the old traditions of the iatrochemists on a basis of sound experimental evidence, as the newer views of scientific methods demanded. He emphasized often that theories were of use only if they were related to what was actually observed in the laboratory, and was careful to establish that the kinds of matter that he postulated were in fact demonstrable experimentally.[73] No doubt the concept that matter was made up of imperceptibly small particles was so well established that it was not felt to be unduly speculative.

At the particulate level, Stahl stressed the part played in chemical reactions by the motion of particles. He also stressed the differences in mobility of the particles of different substances as a cause of differences in their reactivity, and the part played by heat, which he regarded as the motion of particles. However, his most influential contribution to the discussion of affinity was what became known among mid-eighteenth-century French chemists as his theory of *latus*.

Stahl took from Becher the theory that all matter was composed of the principles of air (which very seldom, if ever, formed part of solid compounds), water, and three sorts of earth. These three sorts of earth, none of which was ever found pure in Nature, were the inflammable (phlogiston), the liquefiable or mercurial, and the vitrifiable or fusible. They correspond roughly with the sulphur, mercury, and salt of Paracelsus. Stahl believed that salts were composed of earth and water, a theory which he based on analysis but was unable to confirm by synthesis.[74] He also accepted the traditional view that substances tend to combine because of a similarity in their composition.[75] That led him to the suggestion that an acid attacks a metal because of similarity of composition in the sense that the same earth enters into the composition of both. This aspect of the theory of affinity, which deals with similarity of principles rather than particles, is shown for instance in his *Treatise on Salts*. Nitric acid attacks metals by inflammable earth which they both contain, and also by the mercurial earth which is contained in the metals and is much influenced by the inflammable earth. The acid of marine salt (hydrochloric acid) on the other hand attacks metals by the mercurial earth directly.[76] It is hard to know what precise meaning to attach to the statement that the mercurial earth is influenced by the inflammable earth, which sounds curiously like the kind of allegorical language which was used in alchemical writing, and out of place in the enlightened eighteenth century.

In his *Zymotechnia fundamentalis*, however, Stahl had written rather of the features of the particles of matter themselves which brought about this tendency of substances which were similar in composition at the observable level to combine. He wrote, for instance:

> How does an aqueous corpuscle have the property that it can be brought to bear on fermentable corpuscles (composed of salt, oil and the subtle earth)? We answer: because there is in fact in the corpuscle itself a saline portion, which is composed of an aqueous and an earthy corpuscle, thoroughly and intimately intertwined. Hence indeed the aqueous corpuscle from outside either is easily united through the fact that it is already connected with the terreo-saline one from its very similarity of substance, or of shape and size, or else attacks this whole minute saline corpuscle, as if from one side (*latere*) or the other, and is brought to bear on it, since it is undoubtedly proportioned to this mutual bringing to bear, which this actual combination of an aqueous particle, with which it produces a saline compound, sufficiently demonstrates.[77]

The notion of one particle's attacking a compound particle by the *latus* of the compound particle where there is a component similar to the attacker is explained a little earlier: 'It must therefore be the case that this instrumental mover either impels the integrant molecule itself as a whole, immediately, or else by acting mediately on a single particle, as if on one side, at the same time puts in motion the others which are connected with it.'[78]

One may naturally object that Stahl has given no explanation why a particle should attack another particle of the same kind as itself rather than any other kind of particle. The phrase 'proportioned to this mutual bringing to bear' (*ad hanc applicationem mutuam ... proportionatum*) does not on close inspection seem to mean much. It does not seem to refer to the common suggestion that substances might unite because the particles of one fit into the pores of another. All that Stahl's phrases do is to push the need for an explanation one stage further back. However, the logical status of the theory is not the point in question here. What matters is that it was used by later chemists as a guide to their thinking and experimenting.

In the *Fundamenta chymiae* Stahl gives an even more mechanical, though consistent, description of the means by which salts (i.e. acids, in this case) have the power to dissolve solids:

> Now as Fire, Water and Air are the principal Instruments of Motion; so Earth, of itself, is the Principle of Rest and Aggregation; though by accident, and especially when joined with Water, or Water and Air, in the form of Vapour, it lays the foundation of Separation or common Solution.
>
> For Salts, which are the most subtile terrestrial Concretes, have the mechanical modus of their dissolving power dependent upon this, that being angular bodies, with Surfaces interposed between the Angles, and agitated in a liquor, they force out of their places all the particles of Aggregates; that is, they destroy the connexion of aggregation. And in the sides of terrestrial particles there is lodged an aptitude and disposition for contiguous or aggregative Combination; as there is an aptitude in their points and angles to penetrate, adhere and drive off.[79]

Stahl, then, suggested to chemists in the early and middle eighteenth century various ways in which they might elaborate the primitive idea that chemical combination is due to similarity. However, his theory of *latus*, though it was popular for a while, then faded from view. Part of the reason was that the mechanism which it described was in principle unobservable and therefore it was out of order to mention it. A more general reason was that substances which were not at all similar to each other seemed to combine quite as often as those which were in some respect similar.

It will be evident from the work of other writers who have already been quoted that the general style of Stahl's mechanical and other theories is normal for his time and not unique; but his work was among the most widely read and so among the most influential. At the level where he is discussing observable substances rather than particles he gives a fairly elaborate account of the order of preference of metals for combination with acids and alkalis, more developed than Newton's but crude compared with Geoffroy's table (which will be discussed in Chapter IV.) Stahl wrote in the *Fundamenta chymiae*: 'Those Bodies that are soluble respectively, are thrown down, separated or precipitated out of the Menstruums which dissolve them, by those that will not dissolve them: thus Alkalies precipitate the solutions made by Acids, and vice-versa.'

He enlarges on this statement in more detail:

As a Rule for Precipitation, let the following particulars, relating to the Order in which Metals dissolve, be carefully observed. (1) Gold, Copper and Iron also in part, are dissolved by Aqua Regia, but all metals, except Gold, by Aqua fortis and Spirit of Nitre [both names for concentrated nitric acid]; tho' one more readily than another. (2) Suppose Silver, therefore, dissolved in good clear Aqua-fortis or Spirit of Nitre, and the Clear Solution decanted. (3) If plates of Copper be put into this solution, the Menstruum [solvent] will dissolve them, and let fall the Silver, it had before dissolved, in the form of a white powder. (4) When the Menstruum will dissolve no more Copper, let the clear solution be decanted from the Silver-powder at the bottom, and poured upon Iron-filings; and it will attack these with a considerable noise, dissolve them, and let fall the Copper it before contained. (5) This Solution of Iron being filtered, and poured upon Zink; the Menstruums will dissolve that, and let fall the Iron. (6) Let this clear Solution of Zink be separated from the Iron, and poured upon Crabs-eyes or Oyster-shells [*i.e.* calcium carbonate], and the Menstruums will now dissolve these and let go the Zink. (7) If this Solution be filtered, and Spirit of Urine, or Spirit of Sal-ammoniac [ammonia] be poured thereto; the Menstruum will imbibe this, and let go the testaceous Body [i.e. the calcium salts]. (8) Lastly to this clear Liquor add the Liquor of fixed Alkali [potassium or sodium hydroxide] and the urinous Salt [the ammonia] will be separated; though it does not fall to the bottom, but according to its own nature, rises to the top.[80]

This paragraph conveys much the same information as the column for nitric acid in an affinity table, though in a more cumbersome way. Also, being more specific about the reactions to which it refers, it is less liable to exceptions than the more generalized affinity tables. Stahl repeats the list in a shortened form near the beginning of the section on salts in the *Fundamenta chymiae*.[81] It represents an effort to make observations of chemical phenomena more systematic and precise, but without the sophisticated and perhaps over-elaborate though elegant patterns imposed by affinity tables.

Boerhaave

As we have seen, Stahl was in most ways pre-Newtonian in his attitude to chemistry. Hermann Boerhaave (1668–1738), whose influence on chemistry in the first half of the eighteenth century through his teaching at Leiden was very extensive, shows an intermediate stage of blending Newtonian and pre-Newtonian views. Although like Stahl he criticized his predecessors' over-emphasis on the role of chemistry in medicine, he was primarily a teacher of medicine. His lectures on chemistry (like Stahl's lectures and books) were intended for medical students and were more concerned with practical chemistry than with theory. They were first published in a pirated edition (*Institutiones et experimenta chemiae*, 2 vols, Paris (but in reality probably Leiden), 1724); and Boerhaave was consequently driven to publish his own version (*Elementa chemiae*, 2 vols, Leiden, 1732). Even if they were not what are now called 'iatrochemists', that is believers in the chemical interpretation of physiology, practising physicians generally prepared their own medicines and were concerned with chemistry for that reason. However, Boerhaave's work on chemistry was naturally known to a considerably wider public of experimental philosophers.

His influence was particularly strong in Scotland, which had a close connection with Leiden. Forty Scots professors during the eighteenth century had studied under him. His pupils included James Craufurd, the first professor of chemistry at Edinburgh, and Andrew Plummer, whose lectures were modelled on those of his teacher. Every member of the medical faculty of medicine founded at Edinburgh in 1726 had studied at Leiden.[82]

In medicine, Boerhaave has been described as an iatrophysicist, that is a believer in the appropriateness of mechanical or physical as opposed to chemical or vitalist explanation of bodily phenomena. In this way he differs from Stahl, who in biology was a vitalist. Boerhaave also had the reputation of being a Newtonian. That does not mean that Newton's ideas of chemical attraction were central in his chemical theory, but that he was the first teacher of chemistry to present it as an exact science which might be thought of as part of the Newtonian synthesis. It may also mean, however, that he was Newtonian in his attitude to scientific method. His most notable characteristic is his insistence on experiment, and on exact and careful observation and record, both in medicine and in chemistry; and it was for this that his teaching was valued. He was more an exponent, developer, and critic of the theories of others than an originator of new theoretical lines of his own.

Like virtually all eighteenth-century chemists, Boerhaave accepted that matter consisted of minute particles or corpuscles. However, he still spoke of the four Aristotelian elements, and sometimes of the three principles. For instance, he said that metals were composed of a mercurial base and another, subtler constituent, which he agreed to call sulphur.[83]

His views on the causes of chemical combination, a synthesis of Newton's ideas with earlier mechanical notions, are clearest in the section on menstruums (solvents) of the *Elementa chemiae*. He regards solution as the complete intermingling of the particles of the solute with the particles of the solvent.[84] The particles of each are themselves unchanged, but their combination with each other produces a new set of properties and is not merely a mechanical mixture. Among the factors promoting solution he emphasized motion, which would bring the particles of the solvent and the solute into contact with each other.[85]

Another important factor is Fire. Boerhaave, like Boyle and Galileo, believed that fire was a substance, which he thought to be continually in motion, and continually passing in and out between the particles of other substances, and that its particles repelled each other, thus separating the particles of other substances and causing them to expand. Fire would therefore promote solution by setting the particles of both solute and solvent in motion and also by separating them from other particles of their own kind, which is a necessary preliminary to solution.[86]

Boerhaave classified menstruums into four types. The first consists of those which act by mechanical means also. Here he stresses the mechanical explanation, in terms of the shapes and sizes of the corpuscles. 'There must be some proportion between the Pores of the Body to be dissolved, and the particles of the Menstruum by which the solution is to be performed.'[87] If there was such a proportion, then the particles of the menstruum would be able to get into the pores. Boerhaave's suggestion of a relation between the pores of a solute and the particles of a solvent recalls the ideas of both Boyle and Newton.

His second kind of menstruum is that which acts by mechanical means and in addition by means of repulsion. The kind of repulsion he means is that to which the repulsion between oil and water belongs, but rather surprisingly the example he gives is the result of putting molten copper, silver, or gold into cold water. A strong repulsion divides the metal into very fine particles.[88]

However, the third kind of menstruum is the important one. Where one substance which is much heavier than another remains dissolved in it instead of being precipitated by its weight, some other cause must be found for the motion which brings about the separation of particles of the same kind from each other and their union with particles of a different kind, 'and not rather, that after the solution is completed, the Particles of both of them should, by an affinity of nature, collect themselves into homogeneous Bodies'.[89] This passage has been misunderstood to mean that 'affinity' is among the causes of solution. However, Boerhaave's point is the opposite. One would expect the particles of the solute to cluster together and those of the solvent to cluster together, each separately from the other, according to the common opinion of his time, on account of their respective similarity or affinity of nature; but that is just what they do not do. Some other cause is therefore to be sought. It is here that Boerhaave introduces Newton's concept of chemical attraction. His example is a familiar one, the solution of gold in *aqua regia* [a mixture of nitric and hydrochloric acids], which is eighteen times lighter.

> It is not plain, therefore, that between every particle of the Gold and Aqua Regia there is some reciprocal virtue, by which they attract, and come into a close union with one another? …
>
> If we may be allowed, therefore, to reason from analogy, the action of dissolving, so far as we are hitherto acquainted with it, seems rather to be performed by a certain power, by which the Particles of the *Menstruum* endeavour to associate to themselves those of the Body to be dissolved, than by one which makes them fly from, or repel one another. Here, therefore, we are not to conceive of any mechanical actions, violent repulsions, or natural disagreement, but there seems on the contrary, to be a sociable attraction and tendency towards an intimate union.[90]

One of the main difficulties for those who tried to equate gravitation or other attractions with chemical attraction was that chemical attraction differs according to the nature of the substances concerned, whereas gravitation and the rest were normally supposed to be universal. Boerhaave takes this point. He says that the motion by which menstruums act is not due to 'any of the common origins of motion', for example Propulsion, Gravitation, Electricity, Magnetism; 'for this is peculiar to the Solvent and Solvend, and not common to Bodies in general'.[91]

Therefore, when complete solution takes place in spite of a considerable difference in specific gravity

> we then ought to conclude, that this Solution was effected partly by a universal, mechanical power, which almost always concurs, but chiefly by some other cause

arising from the particular relative disposition of the Solvent and Solvend to one another, by which power we suppose the Elements of one attract those of the other from their former concrete, and being combined together after the Solution is complete, compose numberless new *Species* of Bodies.[92]

Boerhaave's fourth class of menstruum is the largest, the class of those which act by all three means, mechanical means, attraction, and repulsion.[93]

He is, then, quite untroubled by any philosophical objections either to action at a distance, which is implied by the concepts of attraction and repulsion, or to such phrases as 'sociable attraction and tendency' as implying animism, or to speculations about the shapes and sizes of insensibly small particles. He is simply the down-to-earth practical chemist selecting from the various available theories in the way which seemed best to suit the facts. It is plainly this attitude which appealed to his contemporaries so much; and indeed it is an attitude which seems to have been shared by a great many eighteenth-century chemists who did not write on the subject of chemical affinity. It is a symptom of the growing autonomy of chemistry as a discipline. However, in the next chapter we must turn to those who did write about chemical affinity, and who favoured some particular theory.

Notes

1. Empedocles, ap. Aristotle, *Metaphysics*, A4. 985a 21.
2. See for instance Ho Ping-Yü and J. Needham, 'Theory of categories in early medieval chinese alchemy', *Journal of the Warburg and Courtauld Institutes*, **22**, 3–4, 1959, pp. 173–210, especially pp. 195 ff, and J. Needham, *Science and civilization in China*, Vol. V, Cambridge, 1974–.
3. See S. H. Nasr, *Science and civilization in Islam*, Cambridge, Mass., 1968, pp. 252–3.
4. See A. M. Duncan, 'Styles of language and modes of chemical thought', *Ambix*, **28**, 1981, pp. 83–107.
5. Albertus Magnus, *Liber mineralium* (no place or date, but probably printed by Cornelis de Zinkzea at Cologne about 1499), Liber IV, Capitulum V, fol. xcv. Paul Walden, *Journal of Chemical Education*, **31**, 1954, p. 27, quotes this passage from the 1954 Rouen edition.
6. Geber, *Works*, trans. R. Russell, ed. E. J. Holmyard, London, 1928, pp. xii, 181, 192, 194; referred to by Walden, ibid.
7. Paracelsus, *Werke*, ed. Sudhoff, Munich and Berlin, 1928, Vol. 11, p. 365. Referred to by Walden, ibid., where the volume number is given as Vol. II.
8. The section containing the explanation of this reaction is not in the 1610 or 1611 editions of the *Tyrocinium chymicum*; but in the first French edition (Béguin, *Elémens de chymie*, Paris, 1615, pp. 167–8) it is printed as a continuation of the original recipe. That was pointed out by Prof. T. S. Patterson in 'Jean Béguin and his *Tyrocinium chymicum*', *Annals of Science*, 2, 1937, pp. 243–98 (esp. pp. 265 and 177–8). The addition also appears in the 1618 edition

(Béguin, *Tyrocinium chymicum*, no place or date, but probably Frankfurt-an-der-Oder and certainly 1618), the title-page of which says it was translated into Latin by Jeremy Barth, a Silesian doctor. In the 1634 Pelshoefer edition (Béguin, *Tyrocinium chymicum*, Wittenberg, 1634) and later editions this explanation of the preparation of butter of antimony is headed 'Nota Barthii'. However, that is probably due to a misunderstanding. The title-page of the 1618 edition does say that it was translated from French into Latin and enriched with additional notes and preparations by Barth; and in the letters from Béguin to Barth printed in the introduction to that edition Barth is asked to translate the French version into Latin. However, as Prof. Patterson has shown, Barth can have translated only the material which appeared for the first time in the French edition, the rest having already appeared in Latin. It is unlikely that the note in question was originally due to Barth, as it does appear in the French version, and the letters make it clear that Barth had nothing to do with that. Nevertheless, some of the sections headed 'Nota Barthii' in the 1634 Pelshoefer edition had not appeared before 1618 and are probably due to Barth. The Latin versions of this particular note, of which I have quoted in the text a translation from the French version, do have one sensible addition: after the words 'the vitriolic spirit has an extreme sympathy with the metals', they have the words 'except for gold' (*excepto auro*).

9. J. R. Glauber, *Works*, translated by Packe, London, 1688, Part II, pp. 11–12; referred to by Marie Boas, *Robert Boyle and seventeenth-century chemistry*, Cambridge, 1958, pp. 161–2, and by A. R. Hall, *The Scientific Revolution*, London, 1954, p. 317; rev. edn, *The revolution in science 1500–1750*, London, 1983, pp. 274–5.
10. Glauber, *Works*, trans. Packe, London, 1688, Part I, p. 37.
11. O. Tachenius, *Hippocrates chimicus* and *Clavis*, trans. J. W., London, 1677, *Clavis*, pp. 1–2.
12. 'Cum attamen nulla in rebus fiat vicissitudo, aut transmutatio, per somniatum appetitum hyles; sed duntaxat solius fermenti opera': J. B. van Helmont, *Ortus medicinae*, Amsterdam, 1648, p. 111.
13. John Mayow, *Medico-physical works* (translation of the *Tractatus quinque medico-physici*, 1675, published as Alembic Club Reprint No. 17), Edinburgh and London, 1957, reissue edn, p. 36.
14. Mayow, ibid., p. 32.
15. Mayow, ibid., p. 35.
16. Mayow, ibid., p. 16.
17. Mayow, ibid., pp. 24–5.
18. Mayow, ibid., p. 161. Mayow's views on affinity are discussed by J. R. Partington, *History of chemistry*, Vol. II, London, 1961, p. 606.
19. Boyle, *Works*, ed. Thomas Birch, 2nd edn, London, 1772, Vol. IV, p. 289. Quoted by M. M. Pattison Muir, *A history of chemical theories and laws*, New York, 1907, p. 379.
20. Id., *Electricity and magnetism*, Oxford, 1927; a reprint for the British Association of Boyle's 'Experiments and notes about ... electricity', London, 1675, and 'Experiments and notes about ... magnetism', London, 1676.
21. Id., *Works*, abridged by Peter Shaw, London, 1725, Vol. II, pp. 711–26.
22. Id., *Mechanical origin of corrosiveness and corrosibility* (first published 1675) in *Works*, ed. Thomas Birch, London, 1772, Vol. IV, p. 319. Quoted from

the abridged version by Marie Boas, 'The establishment of the mechanical philosophy', *Osiris*, **10**, 1952, p. 480.

23. Boyle, ibid., p. 230.
24. Id., *Works*, ed. Birch, 2nd edn, London, 1772, Vol. IV, p. 285.
25. N. Lemery, *Cours de chymie*, Paris, 1675; Baron's edn, Paris, 1756, e.g. pp. 733–4.
26. Boyle, *Works*, Vol. III, p. 464.
27. On Newton's work in general, see R. S. Westfall, *Never at rest: a biography of Isaac Newton*, Cambridge, 1980; J. Fauvel, R. Flood, M. Shortland, and R. Wilson (eds), *Let Newton be!*, Oxford, 1989; A. Rupert Hall, *Isaac Newton: adventurer in thought*, Oxford, 1989. Among the very extensive literature on Newton's chemistry, alchemy, and related topics, the following are particularly relevant: H. Metzger, *Newton, Stahl, Boerhaave et la doctrine chimique*, Paris, 1930; B. J. T. Dobbs, 'Newton's alchemy and his theory of matter', *Isis*, **73**, 1982, pp. 511–28; eadem, *The foundations of Newton's alchemy: the hunting of the Greene Lyon*, Cambridge, 1975; reviews of that work by K. Figala, 'Newton as alchemist', *History of Science*, **15**, 1977, pp. 102–37, and by M. Boas Hall, *British Journal for the History of Science*, **10**, 1977, pp. 262–4; A. R. Hall, *Philosophers at war: the quarrel between Newton and Leibniz*, Cambridge, 1980; A. R. and M. B. Hall, *Unpublished scientific papers of Isaac Newton*, Cambridge, 1962, pp. 321–31; eidem, 'Newton's chemical experiments', *Archives internationales d'histoire des sciences*, **11**, 1958, pp. 113–52; eidem, 'Newton's theory of matter' *Isis*, **51**, 1960, pp. 131–44; J. E. McGuire, 'Transmutation and immutability: Newton's doctrine of physical qualities', *Ambix*, **14**, 1967, p. 85; id., 'Force, active principles and Newton's invisible realm', *Ambix*, **15**, 1968, pp. 154–208; id., 'The origin of Newton's doctrine of essential qualities', *Centaurus*, **12**, 1968, pp. 233–60; id., 'Atoms and the "analogy of nature": Newton's third rule of philosophizing', *Studies in History and Philosophy of Science*, **1**, 1970, pp. 3–58; id., 'Existence, actuality and necessity: Newton on space and time', *Annals of Science*, **35**, 1978, pp. 463–508; J. E. McGuire and P. M. Rattansi, 'Newton and the "Pipes of Pan" ', *Notes and Records of the Royal Society of London*, **21**, 1966, pp. 108–43; P. M. Rattansi, 'Newton's alchemical studies', in A. G. Debus (ed.), *Science, medicine and society in the Renaissance*, London, 1972, Vol. II, pp. 183–98; id., 'Some evaluations of reason in sixteenth- and seventeenth-century natural philosophy', in M. Teich and R. Young (eds), *Changing perspectives in the history of science*, London, 1973; P. M. Heimann, 'Nature is a perpetual worker', *Ambix*, **20**, 1973, pp. 1–25; A. Thackray, ' "Matter in a nutshell": Newton's *Opticks* and eighteenth-century chemistry', *Ambix*, **15**, 1968, pp. 29–53; R. S. Westfall, *Force in Newton's physics*, London and New York, 1971, esp. pp. 323–423; id., 'The foundations of Newton's philosophy of nature', *British Journal for the History of Science*, **1**, 1962–3, pp. 171–82; id., *Science and religion in seventeenth-century England*, New Haven, 1958; id., 'Newton and the Hermetic tradition', in A. G. Debus (ed.) *Science, medicine, and society in the Renaissance*, London, 1972, pp. 183–98; D. Geoghegan, 'Some indications of Newton's attitude towards alchemy', *Ambix*, **6**, 1957, pp. 102–6; T. S. Kuhn, 'Newton's "31st Query" and the degradation of gold', *Isis*, **42**, 1951, pp. 296–8; id., 'Reply to Marie Boas', *Isis*, **43**, 1952, pp. 123–4; *id.*, 'The independence of density and pore-size in Newton's theory of matter', *Isis*, **43**, 1952, p. 123; A. G. Debus, *The chemical dream of the Renaissance*, Cambridge, 1968. See also F. E. Manuel, *The religion of Isaac Newton*, Oxford, 1974.

28. See for example D. McKie, 'Some notes on Newton's chemical philosophy', *Philosophical Magazine*, **33**, 1942, pp. 847–70; R. S. Westfall, 'Isaac Newton's Index Chemicus', *Ambix*, **22**, 1975, pp. 174–85.
29. John Harris, *Lexicon technicum*, Vol. II, second page of Introduction. See D. McKie, 'John Harris and his *Lexicon technicum*', *Endeavour*, **4**, 1945, pp. 53–7; G. Bowles, 'John Harris and the powers of matter', *Ambix*, **22**, 1975, pp. 21–38.
30. Among standard works see for instance D. Brewster, *Memoirs of ... Sir Isaac Newton*, London and New York, 1965 (repr. of Edinburgh, 1855, edn), 2 vols, Vol. II, pp. 123–56; L. T. More, *Isaac Newton: a biography*, New York, 1962 (repr. of 1st edn, New York, 1934), pp. 380–92; and R. S. Westfall, *Never at rest*, Cambridge, 1980, pp. 533–40.
31. I. Newton, *Mathematical principles of natural philosophy*, trans. A. Motte (1729), revised F. Cajori, Berkeley, 1934, p. 192.
32. A. Thackray, *Atoms and powers*, Cambridge, Mass., 1970, p. 55.
33. Cf. Note 31 above.
34. Newton, *Opticks*, New York, 1952 (repr. of 4th edn, London, 1730, of work first published in 1704), pp. 375–6. See I. B. Cohen, *Franklin and Newton*, Philadelphia, 1956, esp. pp. 164 ff.
35. Newton, ibid., p. 377.
36. Ap. Boyle, *Works*, ed. Birch, London, 1744, Vol. I, p. 71: repr. in *Isaac Newton's papers and letters on natural philosophy*, ed. I. B. Cohen, Cambridge, 1958, p. 251. See also in the same volume the introductory essay by M. Boas, 'Newton's chemical papers', pp. 241–8.
37. Newton, *Opticks*, New York, 1952, pp. 380–1.
38. Newton, ibid., p. 385.
39. Newton, ibid., pp. 382–3.
40. Newton, ibid., p. 388.
41. Loc. cit.
42. Newton, ibid., pp. 354–8.
43. Newton, ibid., p. 388.
44. Newton, ibid., pp. 388–9.
45. Newton, ibid., p. 390.
46. Newton, ibid., p. 394.
47. Loc. cit.
48. Newton, ibid., p. 395.
49. Newton, ibid., p. 396.
50. Newton, ibid., pp. 401–2.
51. Harris, *Lexicon technicum*, Introduction; reprinted in *Isaac Newton's papers and letters on natural philosophy*, ed. Cohen, Cambridge, 1958, p. 258.
52. Harris, *Lexicon technicum*, Introduction (reading 'them' for 'the' as an obvious misprint). See also Boas, 'Newton's chemical papers', in *Isaac Newton's papers and letters on natural philosophy*, ed. Cohen, pp. 245–6.
53. A. R. and M. B. Hall (eds), *Unpublished scientific papers of Isaac Newton*, Cambridge, 1962, pp. 321–31.
54. See A. Koyré, *Newtonian studies*, London, 1965, pp. 139–48, 149–63; E. J. Aiton, *Leibniz: a biography*, Bristol, 1985.
55. Newton, *Four letters from Sir Isaac Newton to Doctor Bentley*, London, 1756, p. 20; reprinted in Hall and Hall (eds) *Unpublished scientific papers of Isaac Newton*, Cambridge, 1962, p. 298.
56. Hall and Hall, ibid., pp. 214–20, esp. pp. 215–16.

57. Jakob Bernoulli, *Werke*, Band I, Basle, 1969, pp. 318–400.
58. Newton, *Correspondence*, ed. H. W. Turnbull, Vol. I, Cambridge, 1959, p. 364.
59. Newton, ibid., p. 365. 'Volatile saltpeter' is presumably a reference to Mayow.
60. Newton, ibid., p. 366.
61. Boyle, *Works*, ed. Birch, London, 1744, Vol. I, p. 70 (*Isaac Newton's papers and letters on natural philosophy*, ed. Cohen, p. 250).
62. Boyle, ibid., pp. 70–1 (Cohen (ed.), ibid., pp. 250–1.)
63. Loc. cit. See also the letter to Oldenburg of 25 January 1675–6, reprinted in Boyle, ibid., p. 74 (Cohen (ed.), ibid., p. 254.)
64. Hall and Hall (eds), *Unpublished scientific papers of Isaac Newton*, Cambridge, 1962, p. 315; discussed pp. 205–7. See also L. Rosenfeld, 'Newton's views on aether and gravitation', *Archives for the History of the Exact Sciences*, **6**, 1969, pp. 29–37.
65. Henry Guerlac, 'Newton's optical aether', *Notes and Records of the Royal Society of London*, **22**, 1967, pp. 45–57.
66. Joan L. Hawes, 'Newton's revival of the aether hypothesis and the explanation of gravitational attraction', *Notes and Records of the Royal Society of London*, **23**, 1968, pp. 200–12.
67. Newton, *Mathematical principles of natural philosophy*, trans. A. Motte, revised F. Cajori, Berkeley, 1934, p. 547.
68. Loc. cit.
69. Newton, *Opticks*, New York, 1952, p. 350.
70. Newton, *ibid.*, p. 352.
71. G. E. Stahl, *Fundamenta chymiae*, Nuremberg, 1723, Praefatio.
72. E.g. Stahl, *Opusculum chymico-physico-medicum*, Halle, 1715, p. 232. Referred to by L. J. M. Coleby, 'Studies in the chemical work of Stahl', unpublished University of London Ph.D. thesis, 1938, p. 20.
73. E.g. Stahl, *Philosophical principles of universal chemistry* (Peter Shaw's translation of the *Fundamenta chymiae*), London, 1730, pp. 3–20. See Hélène Metzger, *Newton, Stahl, Boerhaave et la doctrine chimique*, Paris, 1930, pp. 93–116.
74. Stahl, loc. cit.; *Ausführliche Betrachtung … von den Saltzen*, Halle, 1723, pp. 25–6, 30; *Traité des sels* (trans. by d'Holbach of the *Ausführliche Betrachtung … von den Saltzen*), Paris, 1771, translator's preface, p. iv.
75. E.g. Stahl, *Zymotechnia fundamentalis, seu Fermentationis theoria generalis*, Halle, 1697, pp. 20ff.
76. Stahl, *Traité des sels*, Paris, 1771, pp. 304–5.
77. Stahl, *Zymotechnia fundamentalis*, Halle, 1697, pp. 64–5: '*Unde Aqueum Corpusculum* hoc habeat, ut *Fermentabilibus* Corpusculis (ex Sale, oleo, & terra subtili, compositis) ita applicari possit? Respondemus: Quia jamtum in ipso Corpusculo illo est portio Salina, quae ex *aqueo & terreo* Corpusculis, intime & penitissime complicatis, constat: Unde quidem *extrinsecum aqueum* Corpusculum, vel per illud, quod jam cum terreo-Salino connexum est, ex ipsa *similitudine*, seu *figurae & magnitudinis*, facile conjungitur; Vel ad minimum totum hoc Corpusculum *Salinum, velut ex alio latere*, aggreditur, eique applicatur, cum illud ad hanc applicationem mutuam, utique proportionatum sit, quod actualis jam admixtio *Aqueae* particulae, cum qua *Salinam* Mixtionem exhibet, satis demonstrat.'
78. Stahl, ibid., p. 53: 'Oportet itaque, ut *Movens* hoc *Instrumentale*, vel *immediate* ipsam *integram* moleculam collectim impellat, vel *mediate*, unam particulam,

quasi *unum latus*, afficiendo, reliquas connexas, simul agitet.' It is a slight anachronism to translate Stahl's word *moleculam* as 'molecule', but any other translation would be clumsy: he means by it the smallest assemblage of particles which constitutes a unit of the compound substance as it is observed by the chemist, so that the composition of every integrant molecule of that compound is the same.

79. Stahl, trans. P. Shaw, *Philosophical principles of universal chemistry*, London, 1730, pp. 65–6.
80. Stahl, ibid., pp. 39–40.
81. Stahl, ibid., p. 78.
82. See Archibald Clow, 'Hermann Boerhaave and Scottish chemistry', in Andrew Kent (ed.), *An eighteenth-century lectureship in chemistry*, Glasgow, 1950, pp. 41–8; E. A. Underwood, 'English-speaking medical students at Leyden', *Nature*, Vol. **221**, No. 5183, 1969, pp. 810–14. See also G. Lindeboom, *Hermann Boerhaave. The man and his work*, London, 1968, which, however, does not discuss Boerhaave's chemistry very fully.
83. Boerhaave, *New method of chemistry* (translation by Peter Shaw of the *Elementa chemiae*), London, 1741, Vol. 2, p. 103.
84. Boerhaave, ibid., p. 489. Quoted by J. G. Knight, The chemical studies of Hermann Boerhaave, unpublished University of London M.Sc. dissertation, 1933. (Cf. Dallowe's slightly abridged and corrected translation of the *Elementa chemiae*, *Elements of chemistry*, London, 1735, p. 386.)
85. Boerhaave, trans. Dallowe, *Elements of chemistry*, London, 1735, p. 394.
86. Boerhaave, ibid., p. 396.
87. Boerhaave, ibid., p. 409.
88. Boerhaave, ibid., p. 396.
89. Boerhaave, ibid., p. 391. Boerhaave was not the first to use the term 'affinity' in its chemical sense. Albertus Magnus (*Liber mineralium*, Lib. IV, Cap. V, fol. xcv) had used the Latin equivalent, *affinitas*; and the word had been used in Latin by Mayow, *Medico-physical works*, Edinburgh and London, 1957 (1675), p. 35 *et alibi*; and in English by Boyle, *Works*, ed. Peter Shaw, London, 1725, Vol. II, p. 87; and by J. W., translator of Tachenius, *Hippocrates chemicus*, London, 1677, p. 59 (where the Latin has *sanguine & nativitate*).
90. Boerhaave, trans. Dallowe, *Elements of chemistry*, London, 1735, pp. 391–2.
91. Boerhaave, *op. cit.*, p. 395.
92. Boerhaave, *op. cit.*, pp. 396–7.
93. Loc. cit.

3

Physical theories and their reception by chemists

The needs of chemistry

Eighteenth-century natural philosophers saw their function as finding the reasonable and intelligible patterns that underlay the apparent disorder of the physical world. On the other hand, they were thoroughly convinced that speculations and hypotheses were rash and to be avoided. Any theorizing ought to be a matter of finding regularities in experimentally observed facts. As chemistry developed into an independent branch of philosophy, its practitioners were therefore torn between conflicting demands—they wanted the sense of order and regularity which a general theory would bring, but they were debarred from the most obvious route to such a theory, by way of *a priori* reasoning.

During the century there was a rapid growth in the number of chemical facts known—that is, in knowledge of properties of substances, of methods of preparing them, and of their various reactions, and in the number of different substances that were recognized as being distinct from one another, and hence in the lists of their distinguishing characteristics. This growth in knowledge accompanied a considerable growth in the number of people practising chemistry. It was gradually moving up from the level of alchemy, the apothecary's craft, and the artisan's rule of thumb, and was aspiring to be a respectable science.

The paradigm of a respectable science was Newtonian physics or astronomy, in which everything was neatly docketed and seen to obey simple mathematical laws. The general hope was therefore that chemistry might become like Newtonian physics; but the form which this hope took among practising chemists was that their successors would make their subject like Newtonian physics not by speculation but by codifying the facts, finding patterns and regularities in the facts rather than imposing them from without, and thence inducing mathematical laws. On the other hand, there were many natural philosophers who were not chemists but who graciously undertook to explain, often in the course of explaining much else besides, how the phenomena that were reported from chemists' laboratories could be explained through some variety of the mechanical philosophy, probably more or less closely modelled on

Newton. They were not, of course, proposing these explanations for the benefit of chemists. The intended audience was no doubt the usual audience for enlightened natural philosophy, composed of educated gentlemen of whom few would soil their hands with chemistry. Nevertheless, chemists read of such mechanical explanations and felt obliged to respect them.

The trouble was that such mechanical explanations, though they used concepts and explained aspects of phenomena which were of interest to physicists, did not explain what chemists needed to explain nor describe phenomena in ways which actually assisted the kind of thinking which was native to chemistry. Furthermore, neither the traditional Aristotelian concepts of elements and their interaction nor the systems of chemical principles derived from the three philosophical principles of Paracelsus were any longer appropriate for the needs of chemistry. Although chemists had long been seen, and indeed had seen themselves, as being outside the category of philosophers, and were content to accept such theories as philosophers from more respectable disciplines cared to offer them, that grateful acquiescence no longer satisfied them. They were generally practical men whose business was to get on with finding facts, not at home with mathematics, and cautious about speculative theories which were not soundly based on experiment and observation; yet during the eighteenth century they gradually developed their own homespun concepts and theories which suited the needs of chemistry, and were not imposed from outside it.

At first these concepts and theories were, so to speak, unofficial: lip-service was still paid to the traditional notions and to the doctrines of mechanical philosophers, who at the beginning of the century were still counted as higher in the academic pecking order. The development of chemistry as an autonomous discipline was not, of course, uncontroversial: there were many disputes on the way. Also, our own understanding of that development still presents many unresolved problems. Nevertheless, by the end of the century, through a process which culminated in the work of Lavoisier, the new concepts and theoretical structures which had evolved naturally within chemistry became official and superseded the old models.

In this chapter we shall illustrate the various kinds of system which non-chemists offered to chemistry, the ways in which chemists tried to adapt such systems to the requirements of chemistry, and the growth of autonomous patterns of explanation among chemists.

The commonly accepted view of the proper methods for natural philosophers to use in the mid-eighteenth century is well expressed by Gowin Knight, FRS (1713–72), later the first Principal Librarian of the British Museum. He had made considerable advances in the manufacture of strong artificial magnets, and produced a little book with the title

An Attempt to demonstrate that all the Phaenomena in Nature May be explained by Two simple active Principles, Attraction and Repulsion: Wherein the Attractions of Cohesion, Gravity and Magnetism, are shewn to be one and the same; and the Phaenomena of the latter are more particularly explained.

Knight explains his ideas on scientific method in his Introduction.

Some centuries were spent by the Industrious Schoolmen in Searching for that Knowledge in their own Brains, which was only to be found by a diligent Inquiry into the Works of Nature, careful Observation, and accurate Experiments. Des Cartes was sensible of the Fallibility of Human reason, and imputing to that the Errours of preceeding Philosophers, was resolved to be upon his guard himself. Yet in spite of all his Precautions, the Strength of his Imagination hurried him into many Errours ... [These] Errours taught others how little we ought to rely upon mere Hypothesis, and put them upon a more diligent Application to those rigid tests of Truth, Experience and Mathematics. From this Source we may fairly deduce the many and surprising Discoveries of the last and present Age. Yet even those seem capable of making but little progress, unless rightly applied to the Investigation of the general Laws of Nature; from which (if once known and explained) all the several Effects that depend on them may be readily deduced.[1]

It cannot be said that Knight observed his own advice. In fact, 'the Strength of his Imagination hurled him into many Errours'. Nevertheless, his general view of how truth should be sought, by Experience and Mathematics rather than by reliance on mere Hypothesis, was normal for his time.

Chemists, however, were not expected to be mathematicians. A number of mathematicians and physicists tried to help the chemists by showing how these more exact sciences could be applied to explain the phenomena of chemistry, as has been mentioned already. Chemists did not often take the hint, but they treated such attempts with respect. G. L. Lesage (1724–1803), a mathematical physicist trying to explain chemistry by the principles of mechanics, wrote in the summary of the sixth chapter of his prize-winning *Essai de chymie méchanique*: 'I relegate to this chapter the algebraic calculations, which would have terrified the chemists, if they had found them in the other chapters.'[2]

P.-J. Macquer (1718–84), a chemist of great authority in France in the 1750s and 1760s, wrote in the article 'Gravity' in his *Dictionary of chemistry*:

To compleat this article, we must examine what effects are produced by the gravity of bodies in their combinations and decompositions, that is, in all chemical operations. It is undoubtedly the most important and decisive object for the general theory of chemistry, but it is not within our province, as it cannot well be treated, but by the help of mathematics. In this point these two sciences, which appear so remote from each other, meet. A man sufficiently intelligent, and able in both, might, by treating this matter accurately, throw much light upon it, and lay the foundation for a new physico-mathematical science; or rather, might render the application of algebra and geometry to natural philosophy much more general.[3]

Macquer then goes on to say that the primary and integrant parts of bodies are not visible to the microscope, as the heavenly bodies are to telescopes, so that the application of geometry to the behaviour of such minute particles, on which chemical phenomena depended, was more difficult than in astronomy. Macquer is surely right in suggesting that this relative inaccessibility of the mechanism of chemical reactions is the reason for the comparative slowness of chemistry in becoming an exact and mathematical discipline.

It would be easy to quote a multitude of similar views on the desirability of mathematizing chemistry, but two more will suffice. Tobern Bergman (1735–84), an influential Swedish professor of chemistry, wrote in his *Dissertation on elective attractions*:

In this dissertation I shall endeavour to determine the order of attractions according to their respective force; but a measure of each, which might be expressed in numbers and which would throw great light on the whole of this doctrine, is as yet a desideratum.[4]

Similarly A.-L. de Lavoisier (1743–94) wrote in introducing the column for oxygen which he proposed should be added to the tables of affinities:

What I have just said against tables of affinities in general naturally applies to the one which I am going to present; but I do not think any the less that it can be of some utility, at least until increasing numbers of experiments, and the application of calculation to chemistry, put us in a position to carry our views further. Perhaps, one day, the precision of the data will be brought to the point where the geometer will be able to calculate, in his study, the phenomena of any chemical combination whatever, so to speak in the same manner as he calculates the movement of celestial bodies. The views which M. de Laplace has on this project, and the experiments which we have planned, in accordance with his ideas, for expressing by numbers the force of the affinities of different substances, allow us already not to look on this hope as absolutely a chimera.[5]

Thus it was commonly understood that chemistry should eventually become quantitative, so that the laws governing chemical events could be precisely known and expressed and future events could be predicted exactly. That state of affairs had not yet been achieved, but Lavoisier gives the impression that he thought it was nearer to being achieved than had Bergman or Macquer. Certainly the chemist could no longer be content with the cookery-book chemistry which had been sufficient a century earlier, and in any case he needed to bring some order into the untidy accumulation of facts to make them comprehensible. For those who were tempted to speculate, a wide range of theoretical explanations was available.

Types of explanation

As we have seen, in the seventeenth century 'mechanical explanations' of physical phenomena had become fashionable, that is picturing of the unseen workings of nature as being like visible mechanisms in the manner suggested in the Greek atomic theory. However, since the near-atheism and the insistence on vacua of the atomic theory itself were generally unpalatable, explanations were often expressed in terms of particles or corpuscles rather than atoms. In France the dominant type of theory was Cartesian. The whole of space which might appear to be empty was thought to be full of an ether of imperceptibly small particles, and the vortices of the solar system and the pressures transmitted by the ether, together with the various shapes of the larger particles composing perceptible matter, were used to explain what was observed.

In Britain, however, Newtonianism represented different traditions. Many of Newton's own ideas, especially in the earlier part of his career, owed a good deal to Descartes; but the orthodox Newtonian tradition as it was commonly understood during the eighteenth century (even though many different approaches could fairly be claimed as Newtonian) was different from the Cartesian in several ways. In particular, it was usual to attribute to particles intrinsic powers of attraction and often of repulsion, acting at a distance (which was not, as has been shown, quite what Newton himself asserted); whereas in the Cartesian theory such forces could be transmitted only through the particles of the ether, which touched each other with no intervening spaces. It was not until Voltaire's powerful advocacy that Newtonian physics became generally accepted in France.[6]

Professor Gillispie has argued that in Diderot and Venel there is represented a movement of reaction against the coldly mathematical, particulate philosophy of Newtonian scientists in favour of a more humane, romantic science, and a pre-Newtonian, Stahlian chemistry; and that this is connected with the anti-scientific feelings among the Jacobins which produced the dissolution of the Académie and the execution of Lavoisier.[7] True enough, it is evident that there was such a current of feeling against the success of Newtonianism. It is no doubt related to the kind of opposition to the whole idea of a system of chemical affinity or attraction expressed (as we shall see) by Baron, Monnet, De Fourcy, and Demachy. Professor Gillispie does indeed refer to Demachy, though it should be noted that Demachy himself strongly opposed Stahl's theory of affinity. However, this is only one of many currents of feeling, and there were many chemists who did not share it. It seems to have had little effect on the work of those whose contribution to the progress of chemistry has left its mark.

Affinity and similarity

In Continental works of the early eighteenth century there are still traces of the old notion that similar substances tend to combine. Superficially that was clearly absurd, for substances which are quite different do combine; but the similarity might be said to consist in the two substances' containing a common principle which was responsible for the active properties of both. Stahl, as we have seen, held a theory of that kind. More commonly it was suggested that there was some correspondence between the shapes of the atoms or particles of substances which readily combined, or between the pores in them.

For instance J. L. Clausier wrote in explaining the affinity table which he inserted in his translation of Quincy's *Pharmacopoeia*:

> The reason for the first sequence of affinities, that is, for the similar materials which attach themselves together, when they are at liberty, is the correspondence of their pores, which are equally spaced, and the similarity of their pores; because it is the ethereal matter which brings them together, passing through them in the same manner; and when they are once joined, the passage of this matter being uniform, they are held in this position.[8]

However, for the force which holds such substances together he has a mechanical explanation by the presence of the ether, which he identifies with fire, as it passes through the pores of the matter concerned.[9]

Another theory of this kind was that of J. P. de Limbourg. In 1758 the Academy of Rouen offered a prize for a dissertation on chemical affinities, which was to contain a physico-mechanical explanation of them. Eventually the prize was divided between the authors of two dissertations, of which one, that of Limbourg, was said to be deficient in the mechanical explanation, and the other, that of Lesage, was said to be deficient on the chemical side. This reflects a division at the time between the work of physicists and speculators on mechanical explanations on the one hand and the work of practical chemists on the other. Evidently the Academy of Rouen hoped to reconcile the two points of view.

Limbourg takes the normal view of the importance of determining affinities.

> One understands the importance of knowing these things, not only for determining the most expedient ways of proceeding for analysis and for chemical operations and for deducing the explanation of almost all the phenomena of this Art, but also for determining or conceiving the effects of combinations in Nature and of those in various Arts, such as Pharmaceutical mixtures, tests in Assaying, &c.[10]

As was usual, he begins his dissertation empirically, rather than with any theoretical assertions, by pointing out the observed property of certain substances that are different in nature of tending to approach and unite. This property is called attraction, *rapport*, affinity, or the like. In confining chemical affinity to substances which are different in nature, Limbourg is merely making the conventional distinction between chemical combination and cohesion, which holds together the parts of the same substance. He says that magnetic and electric attraction, cohesion, and gravitation are similar attractions found in physics. It is evident that he is strongly influenced by Newtonian ideas, which was in 1758 still comparatively uncommon among French chemists.

Limbourg distinguishes between the affinity of (i) substances of the same kind, (ii) substances with one component in common, and (iii) different substances, pointing out that not all affinities depend on the identity of two substances or on their having a component in common (the Stahlian notion).[11] He rejects mechanical explanations as equally absurd—'Again, these two systems collapse because they are merely speculative.'[12] The reason why the Academy of Rouen found Limbourg deficient in providing a mechanical explanation is clear: it is not so much that he suggests such an explanation and fails, as that he considers that there is a logical objection to any such explanation. In this he is typical of chemists of his time.

In spite of this logical objection, that is the lack of experimental support, he does proceed to give an explanation of affinity that some purists would have accused of being speculative. Though eighteenth-century chemists agreed that unsupported speculation should be avoided, they disagreed about what counted as speculative, each counting his own theory as being on the right side of the dividing line. Limbourg is clearly torn between the desire not to indulge in theories about things that cannot be investigated experimentally, and dislike of vague and unsatisfactory concepts such as action at a distance, or of the alternative of simply leaving chemical affinity entirely unexplained, which would presumably have been rejected by the Academy of Rouen. The result is less satisfactory than if he had made up his mind one way or the other.

His theory is that affinities depend on facility of contact between the particles of the substances concerned, and on the attraction between them. Contact is not enough in itself: the principal cause of chemical combination is attraction, and it is analogous to the other sorts of attraction found in physics. The causes of all sorts of attraction are obscure, but the explanation of chemical attraction lies in the *analogie des parties* (analogy of particles) between the two substances.

His explanation of *analogie des parties* is as follows:

It is by no means only in the identity of the particles, nor only in the relationship between the pores and the shapes of the particles, as I have already observed ... But it appears that this disposition [i.e. to unite] depends in fact on a number of circumstances, sometimes on the size, the shape, and the arrangement of the particles of the bodies; sometimes on the number, the size, the shape, and the disposition of the pores; sometimes again on the uniformity or concurrence of attractive substances, with or without admixture of repellent particles, which diminish the attraction or the affinity of the others; and sometimes on one or other of these conditions, in different combinations, but which it is hardly possible to determine, except through the effects, because the smallest particles and the pores of the first principles of bodies escape our knowledge.[13]

Limbourg's dissertation is a good example of an uneasy attempt to combine concepts which were appropriate to physics with concepts which were appropriate to chemistry, at least a century before the advent of apparatus and experimental methods which made it possible to link them by actual observations. He has blended features of a number of theories of chemical combination, some already Newtonian, which were then available to chemists; but his idea of 'analogy of particles' is evidently due to inability to escape wholly from the notion that affinity must somehow be due to similarity between the substances involved, or at least to their containing some of the same elements. The dissertation was submitted with the motto *non tam idem eodem, quam similis simili gaudet* (same does not rejoice in same so much as like in like), whereas Lesage's had the motto *simile simili gaudet* (like rejoices in like).

The notion that there must be some sort of similarity or correspondence between a solvent and a solute lingers even in the work of the Dijon chemist L. B. Guyton de Morveau (1737–1816) in the last quarter of the century. Guyton followed Buffon in believing that the attraction between the particles of substances which caused chemical combination obeyed the inverse square law and seemed to be the same as the law of gravitation, and that its effects, when acting between particles very close to each other, would be modified by their shapes and relative positions. Solution would take place when the attraction between the particles of a solvent and those of a solute was stronger than that between the particles of the solute themselves, and so was able to separate them. However, another condition was required for solution—'equiponderance' or exact *rapport* in density between the particles of the solvent and the compound particles of the solute.[14] This suggestion is reminiscent of Limbourg and Clausier, and perhaps even of Stahl.

The Newtonian tradition

The name of Newton was taken in the eighteenth century to support several different types of explanations of chemical events. Newton himself had suggested or hinted at three levels of explanation apart from the purely mathematical, differing in logical rigour and speculative boldness:[15]

1. Through the attractive and repulsive powers of particles linked with their size, arrangement, polarity, hardness, and other properties, though

not usually with their actual shapes, these being unobservable. Some of the earlier interpreters of Newton accepted only attraction and not repulsion. This level of speculation seemed unduly bold to most chemists, though not to all, particularly in the second and third quarters of the century. In the last quarter the objections were less strongly felt.

2. Through the attractive powers, or the attractive and repulsive powers, of the particles of which matter was assumed to consist, the powers being taken as intrinsic properties of the particles without speculation on the nature of the particles themselves. This may be regarded as orthodox Newtonianism. At this level Newton's theories were accepted as common currency by the general run of eighteenth-century chemists from the 1730s on in Britain, and perhaps a generation later in France, though many more were prepared to accept attraction alone than both attraction and repulsion. However, considerable ingenuity was needed to make attraction and repulsion alone explain two aspects which were very important in chemical practice, the selectiveness of chemical affinity and the phenomenon of saturation, and even more ingenuity to explain them with attraction alone.

3. Through the pressure of an ether, a mechanical explanation in fluid dynamics of a mechanical explanation in particle dynamics, which was a little reminiscent of Descartes. Little is heard of speculations of this kind except among Cartesians before the mid-1740s, and they are seldom entertained by practising chemists.

At the first level are the lectures of Dr John Keill, FRS (1671–1721), later Savilian Professor of Astronomy, an example of a mathematical physicist providing explanations of chemical problems. The implication is that mathematical physicists were fully equipped to explain the phenomena of chemistry, and that chemists were not really well enough equipped to do so. The audience addressed by Keill and others like him was that of students of natural philosophy, other natural philosophers, and the general educated public. They did not in general write particularly for the benefit of chemists. However, chemists were evidently expected to take respectful note of them. Keill's lectures were delivered at Oxford from 1700 onwards, and are mainly a geometrical and arithmetical treatment of the dynamics of Newton and Huygens, but include an account of Boyle's calculations of the size of particles of gold, copper, and effluvia from their observed properties.[16] In his introduction Keill attacks philosophers (particularly Cartesians) who claim to give mechanical explanations but substitute speculations about particles which they have never seen, though his own work is open to the same charge.[17] In his eighth lecture Keill discusses the way in which all the properties of matter proceed from the motion, situation, order, figure, and position of their corpuscles. For instance, when salts and metals are dissolved in menstruums (solvents) their parts are separated from each other by fermentation and so set in motion by the fluid surrounding them that they become fluid themselves.[18]

John Freind (1675–1728), who was briefly Professor of Chemistry and a Student (i.e. what in other colleges is called a Fellow) of Christ Church at Oxford before turning to medicine and politics, gave similar explanations of most of the operations of chemistry in his lectures. First he gives nine postulates, of which the following are the most significant parts:

1. That there is an attractive Force, or that all the Parts of Matter are drawn towards one another …

5. That this Force is diffus'd but a very little way; so that when Bodies come to be at some distance, it almost vanishes. Nor does it come to be sensible, unless when the Particles of Matter draw nearer one to the other; But at the point of contact it is the strongest. And therefore the Attractive Force decreases in a Ratio of the increasing Distances, which is more than Duplicate [i.e. more than the square of the distance].

6. This Force is different according to the various Texture and Density of the Particles: But in Gravity 'tis quite otherwise, for that always remains the same, however the Texture of Bodies is chang'd.

7. But the Attractive Force is greater in one side of the same Particle, than in another …

9. The Force by which Particles cohere among themselves arises from Attraction, and is changed in many ways, according to the various quantity of Contact.[19]

A good example of the use made by Freind of his postulates is his attempt to explain why one substance is dissolved with greater ease than another by reference to the size of the pores of the substance and its attractive force. Here his debt to Newton is especially obvious. For instance, spirit of wine (alcohol) easily dissolves rosins, which water cannot, because the very small particles of wine can enter the small interstices between the closely packed particles of the rosin, and the larger particles of water cannot. Gold easily amalgamates with quicksilver (mercury) because the particles of quicksilver are attracted with greater force by the gold than by each other, and are small enough to enter the pores of the metal without difficulty. Silver has the next strongest attractive force, and so is almost as easily amalgamated; but the attractive force of iron and brass hardly exceeds that of quicksilver, and so they are amalgamated only with difficulty unless the attractive force of the particles of the quicksilver for each other is counteracted by some other substance's being mixed with it.

Further on in the same lecture, Freind tries to answer the old question why *aqua fortis* (concentrated nitric acid) dissolves silver and not gold, whereas *aqua regia* (a mixture of nitric and hydrochloric acids) dissolves gold and not silver. He assigns numerical values to the ratios of the sizes of the pores of the two metals, their attractive forces, the size of the particles of the solvents, the cohesion of the silver, the momentum with which the particles of the *aqua fortis* rush against it, and the cohesion of the gold. These figures are chosen merely as possible values, since Freind knows that more exact experiments are needed to establish the exact values; but using the values he has chosen he shows by a complicated piece of arithmetic how the phenomenon could be explained, and finally gives a generalized algebraic treatment.[20]

Freind's theory is obviously open to the charge of being speculative and not founded on experiment, though he himself would no doubt have argued that it followed naturally from the phenomena. In any age, what is counted as being an obvious inference from the results of experiment and what is counted as being wild speculation is very much a matter of convention. Even in Freind's own time, his notions would probably have seemed rather speculative to chemists; but among physicists the influence of Newton was so strong during the period when he was the acknowledged leader of English natural philosophers that theories which so closely followed the suggestions of the *Opticks* would have seemed entirely natural and acceptable. One might indeed object that whereas a general suggestion that this type of explanation was reasonable could be accepted, to go further and suggest which substances had small pores and which had large pores, and even to propose numerical values for unobservable variables, was quite unwarranted. However, in the heyday of what has been called 'the principate of Newton', the prestige of mathematical mechanics was great enough to make Freind's proposals seem satisfying.

In his eighth lecture, 'Of Precipitation', he argues that the shapes of crystals depend on (but are different from) the shapes of the particles which compose them, so that from the shape of a crystal it should be possible for a mathematician to calculate the shape of the particles.[21] This line of thought, which was revived (perhaps independently) by Buffon, found some response among chemists in the latter part of the century.

At the end of the second edition of his book Freind gives a review of it which had appeared in the 'Lipsick' (i.e. Leipzig) 'Transactions', and replies to it. The review attacks the whole Newtonian concept of attraction as hypothetical, which was the normal Cartesian attitude.

Keill soon after published a paper in the *Philosophical Transactions* in which he tried to interpret chemical action in much the same manner as Freind in his lectures.[22] (This was, of course, still within Newton's lifetime.) However, neither Keill nor Freind took up Newton's suggestion that there is a chemical repulsion as well as attraction. One who did was Stephen Hales, who though not primarily a chemist made a considerable contribution to the progress of chemistry through his studies of the quantities of gas evolved from various solids and liquids. Since no distinction was then made between various kinds of gas, as we should now call them, which were all regarded as samples of the element air, Hales made little attempt to test the chemical properties of the samples of gas which he obtained. He assumed that he was simply releasing air which had been absorbed by the solids or liquids. One important step was his insistence that air can lose its elasticity and become 'fixed', even though he did not consider that it became chemically combined with solids or liquids. His explanation of that observation was based on Newton's theories.

Thus he wrote in the Preface to his *Vegetable staticks*, referring to his work on air:

Where it appears by many chymico-statical Experiments, that there is diffus'd thro' all natural, mutually attracting bodies, a large proportion of particles, which, as the first great Author of this important discovery, Sir *Isaac Newton*, observed, are capable of being thrown off from dense bodies by heat or fermentation into a vigorously elastick and permanently repelling state: And also of returning by fermentation, and sometimes without it, into dense bodies: It is by this amphibious property of the air, that the main and principal operations of Nature are carried on; for a mass of mutually attracting particles, without being blended with a due proportion of elastick repelling ones, would in many cases soon coalesce into a sluggish lump. It is by these properties of the particles of matter that he solves the principal Phoenomena of Nature. And Dr *Freind* has from the same principles given a very ingenious *Rationale* of the chief operations in Chymistry. It is therefore of importance to have these very operative properties of natural bodies further ascertained by more Experiments and Observations: And it is with satisfaction that we see them more and more confirmed to us, by every farther enquiry we make; as the following Experiments will plainly prove, by shewing how great the power of the attraction of acid sulphureous particles must be at some little distance from the point of contact, to be able most readily to subdue and fix elastick ethereal particles, which repel with a force superior to vast incumbent pressures: Which particles we find are thereby changed from a strongly repelling, to as strongly an attracting state: And that elasticity is no immutable property of air, is further evident from these Experiments; because it were impossible for such great quantities of it to be confined in the substances of Animals and Vegetables, in an elastick state, without rending their constituent parts with a vast explosion.[23]

Hales clearly recognized the difficulty of explaining the phenomena by a single universal force of attraction such as gravitation. That would merely cause the particles of matter to 'coalesce into a sluggish lump'. Yet matter was in fact observed to change its state of combination with variations in temperature, and also to change chemically in ways which were naturally interpreted by supposing that one substance tended to combine with a second in preference to a third. Thus the forces must vary from substance to substance, and also with temperature, and in addition must cease to act when one substance was saturated with another. All this seemed to require forces of repulsion as well as attraction, and forces which were different between different substances and not the same for all matter.

Hales made use of his theory in the course of his chapter on the Analysis of Air. He there says that the elasticity which causes the great expansion of air generated from the 'fermentation' of peas, wheat, or barley 'is supposed to consist in the active aerial particles, repelling each other with a force, which is reciprocally proportional to their distances'. His language here owes something to Mayow, but the idea is basically Newtonian. Hales quotes from the 31st Query at the end of Newton's *Opticks* the criticism of Boyle's explanation of this repulsion 'by feigning the particles of Air to be springy and ramous, or rolled up like Hoops, or by any other mean than by a repulsive power'.

Hales also uses the mechanism of attraction to account for the replacement of one substance by another in a compound. In describing the generation of 'air' from *aqua regia* during the solution of gold, he writes that the 'air particles regained their elasticity, when the acid spirits which adhered to them were more strongly attracted by the Gold, than by the air particles'.[24]

A number of textbooks of physics in the first half of the eighteenth century expounded Newtonian doctrines with the implication that they could be applied to chemistry, for instance those of Roger Cotes (1682–1716) and Henry Pemberton (1694–1771).[25] However, Pemberton does not introduce Newtonian ideas very much into his posthumously published Gresham lectures, *A course of chemistry*, ed. James Wilson, London, 1771. Perhaps that is because they were intended for people who actually wanted to practise chemistry and might not rise to the level of natural philosophy; whereas his book on Newton's philosophy, published in the year after Newton's death, was intended for a more high-brow audience who might wish to know in principle how chemical phenomena could be explained on Newtonian lines, but would be unlikely to dabble in practical chemistry themselves. Some of the books on Newtonian physics were in English or Dutch; but others were in Latin, and so easily accessible to scholars in other countries, and translations were also made. In particular the school of Leiden was influenced by Newton. W. J. 'sGravesande (1688–1742), who met Newton while he was on a diplomatic mission to England in 1715, and was elected FRS, was appointed professor of mathematics and astronomy on his return to Leiden, where he was of course a colleague of Boerhaave, and wrote a Latin exposition of Newton's physics, based on experiments. It introduces attraction but not repulsion, and was translated into English by J. T. Desaguiliers (1683–1744).[26]

The textbooks of Desaguliers himself and of P. van Musschenbroek

(1692–1761) are not unlike 'sGravesande's. However, after mentioning that magnets repel as well as attract (with a force varying inversely as the cube and a quarter of the distance, he says), Desaguliers goes on to consider other instances of repelling powers in nature. He says that very often bodies which attract one another at certain distances and in certain circumstances, repel each other at other distances and in other circumstances. For instance, the way in which airs and vapours which are evolved in heating or fermentation recede from the bodies which produce them, as soon as they are out of their sphere of attraction, proves the existence of a repulsive force. The example is taken from Hales, no doubt; but the context strongly suggests that it was his interest in magnetism and electricity which caused Desaguliers to adopt Newton's idea of repulsion.[27] Boscovich's more abstract and mathematical work clearly owes much to these earlier expositors of Newton.

Another who followed Newton and Keill was J. Marzucchi in Italy, who explained chemical phenomena on physical principles, through the density, extent of contact, porosity, and mutual attraction of the particles of matter, though without using the concept of repulsion.[28] Boerhaave is of course an instance of a chemist who was influenced by Newtonian ideas but did not adopt them exclusively.

Many other writers explored the various possible ways of interpreting natural phenomena through Newtonian ideas in other fields besides chemistry. In England alone the work of Cheyne, Mead, and James Keill (brother of John) in medicine and physiology was notable, and in speculative or even theological explanations of the nature of matter and its properties that of Purshall, Green, Hutchinson and his followers, Rowning, and Rutherforth.[29] However, the writers whose work has been described above will illustrate sufficiently the range of ideas available to practising chemists.

In spite of this series of Newtonian writings in Britain and Holland, there is little mention of attraction in Germany or France between the work of Stahl and Geoffroy (who does not himself refer to the concept), early in the century, and the 1740s. The almost complete absence of new affinity tables in this period will be discussed in the next chapter. However, the resistance to the idea of attraction, or to uses of the word 'affinity' which resembled attraction, is presumably due to the dominance in France of Cartesianism. That ended only in the later 1730s and early 1740s with the gradual acceptance of Newton's physics.

The *Nouveau cours de chymie*

An exception to that generalization, however, is the *Nouveau cours de chymie suivant les Principes de Newton & de Sthall*, attributed by some to J. B. Senac (1693–1770), which was first published in 1723 with a new edition in 1737. In spite of the spelling of the name in the title, the contents of the book make it clear that the reference is to G. E. Stahl and not to Peter Sthael nor anyone else with a similar name. As the linking of the names of Newton and Stahl, whose theories were considerably different, suggests, it is an elementary work, and does not expound the principles of either very thoroughly. In fact, it is an example of an attempt to apply what were felt to be academically approved theories to chemistry so as to make chemistry also respectable. It is rather eclectic, retaining the Stahlian notion that for a menstruum (solvent) to dissolve a body there must be 'some proportion between the parts of the menstruum and the pores into which they insinuate themselves'.[30] However, it does criticize the Cartesian theory of matter, and it does have some Newtonian ideas in it.

In defining what was later called 'attraction', which the book calls 'Magnetism of Bodies', it cautiously says:

> The disposition which several separated parts have to reunite is called attraction in the books of M. le Chevalier Newton. This term shocks Cartesian ears, but I do not know why. It is used only to represent an unknown cause which brings bodies together. The Philosopher of whom I am speaking also uses the term impulsion, and even says that that is more correct. As there is a variety of opinions on the subject, allow me to make some reflections on it.
>
> In any system, it is necessary to have recourse to a prime Mover. M. Descartes established the circular movement impressed on matter as the principle of all things. From that he produces stars, plants, even animals. Full of a geometrical spirit, this fertile genius sought in a single principle the origin of this sequence of phenomena which matter presents to our eyes.
>
> M. Newton, seeing the difficulties of this Philosophy, sought for another principle, or to speak more correctly instead of applying the moving power to move matter in a circle, he occupies it in pushing bodies one against another; and from this principle he deduces everything which takes place in nature. Those who have attributed occult qualities to this great man should have looked at his principle from this point of view, and they would have found it at least as reasonable as that of Descartes.
>
> But M. Newton goes even further. He believes, or rather he wishes us to suppose, that attraction is an impulsion; but at the same time he tells us that the manner in which it acts is utterly unknown to us, seeing indeed that the weight of bodies is in accordance with their solid mass and not in accordance with their surface. He hints that reason would be almost tempted to believe that no impulsion is capable of producing gravity. What we can be sure of is that if bodies are pushed by another material towards a centre, they are pushed just like a cork towards the surface of water; and it is this which does not agree with any of the systems invented hitherto.[31]

The author then remarks that it is not surprising that Newton is not content with the Cartesian philosophy, for nobody—not even Descartes himself—would regard its physical part as more than possible. He continues:

However that may be, it is certain that there is in nature a magnetism which brings together bodies or their parts. It is to this attractive force that we must attribute the majority of the phenomena which are most surprising in the combination or decomposition of bodies; but before seeking for its cause, let us establish its existence by several examples, and examine its effects.[32]

The author shows a sympathetic understanding of the hesitation shown by Newton in committing himself to any particular theory of the cause of gravitational attraction. There follow four general propositions which give the essence of the theory of Affinity. Examples are given for each.

I. There is between the parts of several substances a certain affinity which makes these parts come close and join themselves to each other.

II. If time is given to these parts to unite slowly, they adopt an arrangement peculiar to each substance (i.e. a particular crystalline form).

III. This affinity or attraction is not equal in all substances.

IV. If two substances are united, and a third supervenes which has more affinity with one of these two substances than that one has with the other, then the one of the pair immediately abandons the other, and goes to join itself to this third substance.[33]

This last proposition is the proposition which Geoffroy had stated in the paper in which he explained his *Table des rapports*, with the wording slightly adapted. (see pp. 115–16 below.) The important difference is that Geoffroy had there avoided using the words 'attraction' and 'affinity', and used the neutral word *rapport* instead. The first and third propositions are also implicit in Geoffroy's views.

The *Nouveau cours de chymie* goes on to summarize in a hundred and five propositions the main reactions of the common acids, alkalis, metals, and neutral salts.[34] These propositions are more general than an affinity table, as they refer often to classes of substances rather than to particular substances, and they give lists of substances which react with some particular substance without necessarily giving the order of their affinity for it. However, they also discuss methods of decomposition and the composition of substances—for instance, salts are said to be made up in their simplest form of a watery part and a subtle earth, when they are fluid, but made solid by the addition of '*de la terre*' ('some earth'). Here the author is following Stahl.

The next section deals with the cause of the 'magnetism' between substances, which is the name used by the author to refer to chemical attraction as well as magnetism in the normal sense. He rejects theories such as Descartes' that gravitation and other types of attraction are due to the impulsion of some such thing as an ether pushing inwards from outside the objects attracted together, accepts the possibility that they may be due to matter's being impelled to take the place of something moving outwards from the centre of the earth (another notion which Newton had favoured very early in his career), but concludes that it is in fact impossible to discover the cause of magnetism. We can discover only the laws according to which it acts.[35] After a discussion of solution, the author proceeds to expound the application of Newton's principles to chemistry by repeating Freind's propositions about attractive forces, and Freind's explanation of chemical processes by means of them.[36]

The context of this intriguing though crude work is still mysterious. Macquer, Fourcroy, Weigel, and Gmelin, and later Kopp attributed it to Senac, who was physician to Louis XIV.[37] Baron, more nearly a contemporary but not quite, said that it was produced from students' notes on lectures given by Geoffroy and Boulduc.[38] If so, Geoffroy's paper in the *Mémoires de l'Académie* of 1718 about his *Table des rapports* represents a remarkably successful attempt to suppress his Newtonianism.[39] Dr Thackray indeed suggests that Fontenelle's comments in the relevant part of the *Histoire de l'Académie* were 'written with this fact in mind'.[40] However, Geoffroy's posthumously published lectures show no sign of the Newtonian theorizing which is prominent in the *Nouveau cours de chymie*, and Baron, who after all was writing a generation later, is probably wrong. As Dr Smeaton has pointed out, the lectures explain the combination of acid and alkali rather as Lemery had done, in the manner of Descartes, by suggesting that the spiky particles of alkali joined together in globules with pores in them, into which the points on the particles of acids could enter and so form a neutral salt.[41] In fact, the symbol for acid spirits in general which Geoffroy used in his affinity table seems to represent a particle with spikes on it. It also seems unlikely that Boulduc, who was Geoffroy's demonstrator, was responsible for the theoretical part of the book. Professor Partington points out that the first person singular, as in lectures, is often used in it; but that is not evidence that Geoffroy was the author. The catalogue of Geoffroy's library, said to have been drawn up by himself although it was published after his death, includes a copy of the *Nouveau cours de chymie* and gives Senac as the author. That is surely good contemporary evidence. The catalogue does not include Newton's *Principia*, but does include the *Opticks* (in English), as well as works by Freind and both Keills, and standard works on chemistry such as those by Becher, Stahl, and Boerhaave.[42]

It is true that Geoffroy had visited England, had been elected a Fellow of the Royal Society, and had acted as a link between that Society and the *Académie Royale*, by means of his correspondence with Hans Sloane. We also know that in 1706–7 he read to the *Académie* his own translation of extracts from Newton's *Opticks*, which did not then include the Queries most relevant to chemical attraction.[43] That was evidently the edition which was included in his library. By the time his Table was published in 1718 he had probably read the later edition of the *Opticks*. As we shall see, however, Geoffroy's *Table des rapports* carefully avoids speculation about the causes of chemical combination, and uses words which do not imply either affinity or attraction. The most likely conclusion therefore is that although Geoffroy knew of Newton's work (including by 1718 probably the chemical Queries), and may have been influenced by it in framing his own view on *rapports*, he was not an avowed Newtonian and did not produce the *Nouveau cours de chymie*. The attribution to Senac is at least possible: we do not know enough about him to be sure.

Venel wrote in the *Encyclopédie* in 1753 that the book 'brought us Stahlianism, and caused the same revolution in our chemistry, which the reflections on attraction published by M. Maupertuis … brought about in our physics, in causing us to accept Newtonianism'.[44] This is an exaggeration both of Maupertuis's role and of the influence of the *Nouveau cours de chymie*. Yet the publication of the book in 1723 does give evidence of an interest in France in applying Newtonian principles to chemistry in what would otherwise appear to have been a period of Cartesian dominance. Certainly, Montpellier is a long way from Paris, then the centre of fashionable ideas; and authors in Montpellier, the seat of an ancient university with a strong tradition in medicine and pharmacy, might have been relatively independent of Parisian influence.

As we have seen there was already some Newtonian influence felt in the work of Boerhaave. The acceptance of Newtonianism is complete in the papers of 1738–9 published in *Essays and observations* by Boerhaave's pupil Andrew Plummer, Professor of Chemistry at Edinburgh, though it is not apparent in the surviving manuscript notes of his lectures.[45] They are chiefly concerned with the preparation of medicines. The following passage from the second of the two papers is characteristic of their Newtonianism.

When spirit of nitre of *aqua fortis* [concentrated nitric acid] is poured upon pieces of silver, the appearances plainly point out that there is an attraction betwixt the particles of the solvent and the silver; and that there is likewise a repulsion of the particles already dissolved from the silver, to make way for the application of new particles of the solvent to other parts of the silver: … The attractive force between the saline particles of the spirit of nitre and the silver must be superior to the power by which the particles of silver attract one another; for it dissolves their cohesion, separates them, and keeps them suspended, in a fluid which has only about one tenth part of the specific gravity of silver … That this power of attraction between the particles of spirit of nitre and those of silver must be very great, will appear, if we consider first, that it requires a very great heat to melt silver; but this melting is nothing else than, by the force of fire, to overcome the cohesion among the particles of silver …[46]

Boscovich, Knight, Baumé, and Higgins

Books which deal with chemistry in the Newtonian tradition are much more common from the 1750s onward. However, there is still a clear distinction between the approach of mathematical physicists explaining how chemistry can be interpreted mechanically and the approach of chemists who are mainly concerned with purely chemical problems but are prepared to try mechanical explanations where they seem likely to be helpful. An example of a mathematical and theoretical development of Newton's ideas at the second level, taking into account only attraction and repulsion and not attributing any shapes or other properties to atoms, is the well-known work of the Jesuit priest Roger Boscovich (1711–87), first published in 1758. It represents the climax of the series of interpretations at the second level which we have already discussed. According to Boscovich all matter consists of indivisible and unextended point atoms, which exert forces of attraction and repulsion varying with distance in both sense and magnitude, so that two atoms attract each other at some distances and repel at others. This law of variation is the same in all atoms, and the attraction and repulsion are therefore mutual.

The force is an attraction at all appreciable distances, inversely proportional in magnitude, as nearly as can be detected, to the square of the distance, and known as gravitation. At insensibly small distances, not more than a thousandth or a fifteen-hundredth of an inch, there are many alternations of attraction and repulsion as the distance between the atoms changes, illustrated by a graph of distance against force at the back of the book. There are many distances, therefore, at which two atoms will neither attract nor repel, but are in equilibrium. When the distance between two atoms tends towards zero, the force of repulsion between them tends towards infinity, so that they can never be mathematically in contact. What appears to be a piece of extended matter is thus simply the sphere of intense repulsion round a point centre.[47]

Those according to Boscovich are all the essential properties of atoms, and together they provide the immediate efficient cause of all the sensible properties of matter. He proceeds to apply this theory to explain those properties, including gravity, cohesion, fluidity, viscosity, elasticity, taste and smell, heat and cold, electricity, and certain chemical operations. Two or more atoms may be so arranged as to constitute a particle of the first order. Two or more such particles may constitute a particle of the second order, and so on. (Here Boscovich is echoing the teaching of Stahl.) A particle composed of two atoms is almost without action on another atom along one direction, but acts strongly on it along another direction, like a magnet. Hence particles with such polarity will cluster together in one way in preference to another.[48] This is an effort to explain the selectiveness of chemical combination, and the theory of positions of equilibrium between repulsion and attraction is an ingenious attempt to explain the stability of chemical compounds. In recognizing that any theory had to explain such phenomena Boscovich is superior to most of the mathematicians and physicists who proposed to supply theoretical foundations for chemistry. In both points, indeed, he is following up hints to be found in the 31st Query in Newton's *Opticks*. However newton had merely suggested that 'As in Algebra, where affirmative Quantities vanish and cease, there negative ones begin; so in Mechanics, where Attraction ceases, there a repulsive Virtue ought to begin'—that is, a linear progression from attraction to repulsion, with a single turning-point.[49] Boscovich's notion of a series of alternations between attraction and repulsion is his own addition to the Newtonian doctrine.

The opposition between attraction and repulsion is treated in even more detail by Knight and Higgins. Knight's book is set out in the form of a mathematical treatise, rather like Newton's *Principia*, with ninety-one numbered propositions and their alleged proofs, corollaries, an axiom, and definitions. Having demonstrated from observed phenomena that attraction and repulsion exist, he argues that since the same substance cannot attract and repel at the same time, there must be two kinds of matter, of which one attracts and the other repels.[50] The particles of matter which mutually repel each other seem also to be subject to the general laws of attraction as far as other matter is concerned. But repulsion and attraction are constant, immutable, and irresistible by any finite force.[51] All primary particles of matter are originally of the same size, and are round—the simplest possible supposition. A number of particles of equal magnitudes, and perfectly round, which attract each other, tend to form into a shape as nearly spherical as possible, so as to get as near as possible to their common centre of gravity.

Every corpuscle of attracting matter will have round it, either adhering to its surface or condensed in the form of an atmosphere, as many repellent particles as will just balance its attracting force; except that in the atmosphere there are corpuscles with an excess of repellent matter which repel each other and so produce elasticity.[52] Since all the corpuscles of attracting matter are surrounded with repellent particles, there can never be any real contact between such corpuscles or bodies compounded of them. When we speak of the contact of corpuscles or bodies, says Knight, no more is meant than that they are brought as near as the repulsion of their surfaces will admit.[53]

All the various degrees of cohesion that are found among bodies are attributed to this one, simple, uniform kind of attraction and repulsion, and to the different sizes of their constituent corpuscles.[54] Knight proceeds to apply this theory to explain the properties of air, fluids in general, water, the universal acid, phlogiston, earths, and so on. The size of the particles has to be considered, as this affects the distance apart of their centres—there is a similarity here to the later ideas of Buffon.[55] The theory does not affect the laws of light, if light is considered as vibrations in a repellent fluid (which was not generally the Newtonian view.) Heat is also considered as the vibration of a great number of repellent particles.

Magnetism is explained at great length in terms of the circulation of a repellent fluid—again a rather pre-Newtonian view. (This was just before the publication in 1751 of Michell's *Treatise of artificial magnets*, in which he suggested that magnetism obeyed the inverse square law.[56]) Knight adds that electricity is explicable in similar terms, though that is too long a story for him to discuss at the moment. Although Boscovich differs from Knight in believing that atoms have no magnitude, and in his use of the notion of the polarity of complex particles, the similarity of many of his ideas to Knight's—for instance in their denial of the possibility of contact between the particles of matter—is so close that direct influence seems evident. The common source of their theories in Newton is not enough to explain their similarities.

The first chemist to use the concept of repulsion, as opposed to physicists condescending to explain the fundamentals of chemistry, was apparently Plummer in Edinburgh.[57] Cullen, as J. R. R. Christie has shown, also used the concept of repulsion of the particles of an ether as a factor in chemical reactions. However, it seems to have been a former pupil of Geoffroy's, Antoine Baumé (1728–1804), who introduced the idea of repulsion into French chemistry.[58] His use of the concept provides a good example of the difference between physicists and chemists. Chemists were concerned with finding patterns and explanations for what they observed experimentally. Newtonian theories might be useful for that purpose; but they were less likely than the more philosophical physicists or mathematicians to be interested in the detailed mechanisms of attraction and repulsion or in imaginative descriptions of the nature of invisibly small particles for their own sake. Most chemists, if they accepted the concept of chemical attraction at all, accepted the likelihood that it was related to gravity in some way, but considered that the complexities of chemistry made the link too difficult to discover, and that perhaps the link was not a simple or direct one. The physicists or mathematicians, on the other hand, were happy to rise above the complexities of chemistry. They made an optimistic assumption that the unseen workings of nature followed tidy, mechanical, material patterns, capable ultimately of being expressed quantitatively in the Newtonian manner, though there was as we have seen some discrepancy over what was the authentic Newtonian manner, and how much theoretical speculation could be justified by, or even based on, the observable phenomena.

The attitude of Macquer, with whom Baumé worked for twenty-five years as demonstrator in his lectures, is fairly typical of chemists, though less cautious than that of many. In the article *Pesanteur* in his chemical dictionary he suggests that gravitation and chemical affinity are due to the same fundamental property of matter, into the mechanism of which he will attempt to make no enquiry; but gravitation on the large scale can be observed and dealt with by physicists, whereas the attraction between the unobservable particles with which chemists are concerned cannot be treated quantitatively—at least, not yet.[59] Guyton de Morveau, towards the end of the century, goes a little further than most chemists towards identifying gravitation and chemical affinity, quoting with approval Buffon's view that chemical attraction at short ranges is probably subject to the same law as gravity but much affected by the shapes of the particles concerned.[60]

Baumé committed himself rather further than his master Macquer on the connection between gravitation and chemical attraction. He wrote in his textbook of 1773:

> The cause which is opposed to absolute attraction, has been recognised by the Physicists with regard to celestial bodies; they have given it the name of repulsion, that is to say, a force quite as real as attraction, which repels bodies, after they have drawn together up to a certain point, and which prevents them from joining. It appears that it is from these two forces, attraction and repulsion, well ordered, that there results the balance and the perfect harmony which it has pleased the Creator to establish in the universe. This repulsion, though a secondary cause, and subject to the laws of attraction, is a property inherent in matter; this property acts conjointly with attraction, down to the elements of substances, and all the operations of chemistry. It even appears that it is from these two effects, in combination, and from their different degrees of action, that there results the variety which one observes in the hardness and density of substances: this repulsion has no less an influence on the phenomena of their combination and separation.[61]

Although Baumé quotes the opinion of physicists with proper respect, he clearly makes an implicit distinction between what they can tell him and what is the proper territory of a chemist. By 1773 chemistry had begun to assert itself as an autonomous discipline of independent standing.

His confidence in the existence of repulsion was based on his belief that he had established it experimentally, as he explained in a curious little paper which he had published in the *Avant-Coureur* in the previous year, 1772. It is entitled 'Réfléxions sur l'attraction et la répulsion qui se manifestent dans la cristallisation des sels'. It recalls Newton's suggestion in Query 31 that in physics a repulsive force should begin where positive ones leave off. Newton had not been able to demonstrate repulsion with evidence as strong as that for attraction, but Baumé claims to have done so.

On leaving a glass cucurbit containing a solution of Glauber's salt (sodium sulphate) at the point of crystallization near a bottle also full of Glauber's salt he found that crystals formed in the solution only on the side of the cucurbit nearest to the bottle; but when a bottle of salt of tartar (potassium carbonate) instead of Glauber's salt was put near the cucurbit of solution, the crystals formed only on the side opposite to the bottle. Baumé suggested that the effect was due to attraction in the former case and repulsion in the latter.[62] Unfortunately that suggestion was refuted by Lavoisier, who showed that these effects were due to the cooling of the walls of the vessels.[63] Presumably Baumé's book had already gone to press. However, the revival of the idea of repulsion in Paris at that time is striking. Perhaps Baumé had been reading Boscovich or Knight.

The contrast between framers like Knight or Boscovich of bold, sweeping hypotheses to explain not just a few properties of matter but every property, and the caution shown by the general run of chemists in avoiding such mechanical speculations, is sharp. One of the few chemists, albeit an Irish one, who did speculate freely was Bryan Higgins (1737 or 1741–1818), who ran courses of instruction in chemistry in his laboratory in Soho. Nevertheless, it is evident that no other chemists took any notice of Higgins's system or followed it up. It was his nephew and assistant William Higgins who afterwards claimed, rather superficially, to have anticipated Dalton.[64] Bryan Higgins published in 1776 the first volume of *A philosophical essay concerning light*, which is almost entirely concerned with chemistry. The later volumes, which would presumably have been about light, never appeared. The first volume also contains a reprint of the syllabus of his chemistry course, which is tantalizing because it tells the reader that a number of surprising things are to be established by experiment—but does not say how.

Higgins taught that there were seven elements—Earth, Water, Acid, Alcali, Air, Phlogiston, and Light. That style of making an individual selection of elements or principles would have been characteristic of the beginning of the century. He then stated that the atoms of Earth and Water attract atoms of the same kind, and the atoms of the other five elements repel atoms of the same kind; but the atoms of Earth, Water, Alcali, Acid, and Air each attract the atoms of the others. Only the atoms of Light repel the atoms of all other elements except Phlogiston, and the atoms of Phlogiston do not attract those of pure Water.[65] The attraction between atoms is more forcible in one direction or axis of each atom than in any other direction, and there is a polarity in all matter. There is only one cause of all these attractions, which in all cases is affected by distance and polarity. The attractions of bodies usually enumerated as distinct properties of matter or laws of nature are nothing more than the sum of the attractions of their elementary atoms, or these forces counteracted in a certain degree by repellent atoms, or these forces exerted to the greatest advantage in bodies whose primary attractions are the strongest, and whose atoms are also arranged in polar order. Gravitation, for instance, is not regarded as a law of nature, but as the sum of the forces of the simple attractions of the atoms of earth for each other and for the atoms of other elements. Phlogiston does not gravitate, possibly because the sum of its repulsions exceeds the sum of its attractions.[66]

Fire, which is a compound of Phlogiston and Light, pervades all known bodies and by its repulsions causes them to expand.[67] It tends to counteract the force of attraction. Inflammable air (the term then used for what is now called hydrogen) is a compound of Phlogiston and Acid.[68] In terms of the phlogiston theory, that would of course explain the evolution of hydrogen when an acid acts on a metal, leaving what was then called the calx of the metal (but is now called the oxide) combined with part of the acid to form a neutral salt, since the pure metal was thought to be a compound of phlogiston with the calx. Light itself, according to Higgins, is all-pervasive; but visible Light, or illumination, is Light in motion.[69] Darkness is Light at rest and so unable to affect our eyes. In fact, illumination and darkness are with respect to Light what sound and stillness are with respect to air. (In short, Higgins's Light is what other people often meant by Ether.) Similarly, heat is merely our sense of certain effects of Fire, and not a substance, so that the phrase 'latent heat' (a reference to Joseph Black's theory) is meaningless.[70] Degrees of heat are as the quantity of Fire in the body or spaces. Cold is privation of Fire.

Higgins emphasizes that the different kinds of atom do not all attract with the same force. In the Syllabus he promises for each element an 'Experimental estimation of the force of this attraction in arrangement of the atoms', and an 'Experimental estimation of this attraction in contact compared with its force at given distances of the atoms'; but he does not say how these estimations are to be performed.[71] He believes that the force of attraction varies inversely as some power of the distance, but he is not certain of the actual power. The Newtonian view that the force varies inversely as the square of the distance is rejected, as well as Newton's other suggestion that it varies as some higher power.[72] Although Higgins quotes Newton frequently and with great respect, he differs from him on these and some other points. For instance, he disagrees with Newton's suggestion in the *Opticks* that particles which attract at small distances repel at greater distances, and that this is the reason why a particle can combine with only a certain number of others before saturation. According to Higgins, each particle either always repels or always attracts each other particle (except for the constant neutrality of Phlogiston and Water). This is the basis of his own theory of chemical saturation, which indeed is the main feature of the *Essay concerning light*, and will be discussed in Chapter V. He denies the existence of a Newtonian ether.[73] Also, he emphasizes that his forces of attraction and repulsion extend to indefinite distances and are never suspended, but merely counteracted by opposing forces. According to Higgins, Newton did not mean that chemical attraction when unopposed reaches only to small distances, and the belief that he did mean that is based on a misinterpretation of the words in the 31st Query in the *Opticks*: 'Since metals dissolved in acids attract but a small quantity of the acid, their attractive force can reach but to a small distance from them.'[74]

The system of attraction and repulsion of atoms which Higgins has proposed is then applied in detail to explain the chemical and some of the physical properties of matter. Matter can take on solid, liquid, or elastic (i.e. gaseous) form according to the net effect of the repulsions and attractions of its atoms. That is, if the attractions are very much stronger than the repulsions, the substance in question is solid; if the attractions are not strong enough to counteract the repulsions at all, the substance is a gas. Phenomena are interpreted entirely in terms of the attraction, repulsion, polarity, and occasionally motion and arrangement of atoms, and the properties of atoms are decided by deduction from phenomena. Higgins insists that the properties of a compound are different from the properties of its constituent parts—an important point which was generally grasped by British chemists, but not yet by Continental chemists. However, he makes it clear that the constituent parts are not transmuted but simply have their properties modified by the interaction between their attractions and repulsions.[75] For instance, he points out that a compound gas or solid may have a specific gravity greater or less than that of either of its components.[76]

Other Continental views

Buffon (1707–88), whose influence has already been mentioned, believed that the fundamental law governing chemical affinity was the same as that governing gravitation between large bodies at considerable distances; but he suggested that for very small particles at very short distances the shape of the bodies would modify the effect of the inverse square law. Hence he predicted that future generations would be able to establish the effective law of attraction between particular substances, and so the actual shapes of their particles.[77] Clearly by this time (1765) Newtonianism had become thoroughly respectable even in France. Buffon's suggestion recalls the ideas of Freind and Knight, though he is more likely to have been influenced directly by Limbourg. Guyton de Morveau quotes Buffon's scheme with approval, presumably because of the pleasing prospect which it seemed to offer of quantifying chemistry; but in practice it proved impossible to apply.[78] However, the same notion is mentioned by Bergman.[79]

Lesage, who shared the prize offered by the Academy of Rouen in 1758 with Limbourg, provides an explanation at the third level of Newtonianism, invoking something very like an ether. He supposed that space was filled with particles in perpetual motion. The pressure of these on a body would be greater on the side facing away from another body than on that sheltered by another body, and so the two bodies would be impelled together with a resultant force inversely proportional to their distance apart. This is, of course, like the theory which Newton had suggested in his Letter to Boyle, which had by now been published. Lesage then explained elective affinity as well as gravitation by assuming variations in the shape, size, and position of pores in various substances and so in the resistance which they offered to the passage of bombarding particles.[80]

Lesage also suggested a complicated method of expressing the inequality of affinities and the consequent reactions.[81] It is one of the fairly uncommon examples of such symbolic representation of chemical reactions in the eighteenth century. He recommends borrowing the sign of conjunction, ☌ , from the astronomers, and the signs for equality, =, 'greater than', >, and 'less than', <, from the algebraists. The fact that fixed alkali unites with acids more strongly than water joins with sulphur would be expressed:

The third symbol in this expression is that used by Geoffroy in 1718 for 'acid spirits', that is acids in general, and was rather out of date among chemists by 1758. The last symbol, which Lesage uses for 'sulphurs', is (but for the absence of a cross-stroke on the vertical line at the bottom, which may be an oversight), that generally used for phlogiston, made up of the symbol for ordinary sulphur with the three little circles (representing drops) which constitute the symbol for oil at the corners.

The earlier, less mathematical and more chemical theory of Clausier, described above (p. 70), is another example of this third type. Yet another is that of Demachy, whose criticism of theories of affinity will be mentioned below. Although he severely criticized tables of affinity because they implied unjustified assumptions about the causes of chemical combination, Demachy's own suggestions seem even more open to the accusation of being speculative. He is, in fact, an unusual example of a chemist who attempted a mechanical explanation of the sort favoured by physicists, with not altogether happy results. He argued that the annual and diurnal motions of the earth's atmosphere, as it revolved with the earth, exerted a frictional pressure on all bodies separated artificially or naturally from the surface of the globe. This pressure, combined with the effects of any obstacles offered to this regular movement of the atmosphere, was the cause of the impulsion which caused substances to combine chemically. Different substances had different tendencies to combine because of their different mobilities, and consequently different responses to impulsion.[82]

All these types of explanation of chemical properties were available in the mid-eighteenth century, and must be seen as the background for the ideas of the more successful chemists. If we except Baumé, few chemists of much importance went far with such speculations, though this caution diminished in the last quarter of the century. They were willing to use the words 'attraction' and 'affinity', but not to weave hypotheses about the mechanisms, presumably to be sought at atomic level, which caused such tendencies. In general, among those who referred to such speculations, the mathematicians tended to identify chemical forces of attraction with gravitation, while physicists saw them as similar to gravitation, but often distinguished capillary attraction, cohesion, and chemical attraction from gravitation. Those chemists who concerned themselves with such things, on the other hand, saw the difficulty of accounting for the selectiveness of chemical attraction if it were identified with gravitation, and considered whether other properties of the particles could be brought into the reckoning to modify the effects of attraction. Comparatively few natural philosophers of any kind considered any connection between chemical attraction and magnetic and electrical attraction, though from the 1730s onwards many considered the possibility of repulsive forces' modifying the effects of attraction. However, if repulsive powers were assumed, they were often attributed to the matter of heat or to elastic fluids (i.e. to what would now be called gases) rather than to the particles of solids or liquids.

Nevertheless, the application of Newtonian ideas of attraction between particles to the explanation of chemical events had a profound influence on chemists, even if they avoided being drawn into speculation.[83] Newtonian explanations assumed that all the relevant factors were mechanical and material, that there was no place for immaterial or supernatural essences or principles, but that matter was made of different kinds of particles which were different because of permanent material differences in their shapes or sizes or their determinate powers to exert measurable forces, and that a mental picture of such particles combining together just like lumps of material of sensible size, even though the particles were in reality insensibly small, could be a correct and realistic picture. In their attempts to find orderly, rational patterns in chemical phenomena, chemists were greatly helped by the feeling that such Newtonian explanations were permissible. This helped them to get away from hampering concepts of vague and perhaps changeable principles and to progress towards the notion of chemical elements which formed the basis of nineteenth-century chemistry. Specifically, it must have helped them to conceive of particles, and so the elements made up of particles, as having a distinctive mass or weight which was conserved through chemical changes. It also helped them to appreciate that the properties of a compound might be quite different from those of the constituent elements, and not merely a blend of them. For the properties of the compound would be secondary properties, resulting from the primary properties of its component particles; and the component particles of a compound, it was generally agreed, would each be a cluster of elementary particles and would have primary properties quite different from those which the elementary particles composing them would possess in their uncombined state. Furthermore, the picture of compounds' being made up of clusters of elementary particles implied that the compounds had a constant composition by weight, and might very well have other unalterable properties such as powers of chemical attraction or affinity.

On the other hand, the Newtonian style of explanation helped chemists to avoid what by the mid-eighteenth century seemed the crude and naïve extremes of Cartesian materialism, as found for instance in Lemery or even Geoffroy's lectures, which explained chemical properties by unverifiable assumptions about the shapes of the particles themselves—for instance, that acids were acid because their particles had sharp spikes. It was disreputable in the eighteenth century to indulge in speculations which went so far beyond anything observable.

Thus Newtonian mechanical explanations were fruitful in eighteenth-century chemistry. Yet they were not sufficient. They were essentially of a kind which suited the requirements of physicists. As chemistry developed into an autonomous discipline, it also developed its own requirements. It needed concepts and kinds of explanation of a different style, to suit its own problems and methods, which were not the same as those of physics. Having, then, examined the mechanical explanations and speculations of which chemists must have been aware, we shall turn to the evolution of these peculiarly chemical concepts. They will be considered particularly as they appear in tables of chemical affinity and attraction and in attempts to classify and quantify affinity and attraction and even to discover laws governing them. First, however, we shall discuss some of the criticisms which were made by chemists of the whole idea of attraction or affinity, which generally represent objections to the importation into chemistry of systems which were seen as appropriate rather in physics; then the definitions of those concepts offered by chemists, which represent adaptations of the physicists' concepts to chemistry; and then the ways in which the words were used, which again show the replacement of alien concepts by concepts which were intrinsically chemical.

Critics of affinity

A prominent critic of the concept of constant chemical affinities was T. Baron d'Hénouville (1715–68), who succeeded his teacher Rouelle as demonstrator at the Jardin du Roi in Paris. In a note in his edition, published in 1756, of Lemery's *Cours de chymie* of 1675, after criticizing Lemery's mechanical explanation of the cupellation of silver, Baron adds:

Thence it follows that without seeking to explain the mechanism of all these precipitations, it is much better to be content with the fact, and at least to avoid saying with the Chymists of today that these precipitations depend on the different degrees of affinities which all these substances have with the same solvent.[84]

Further on, in a note on Lemery's mechanical explanation of the precipitation by sea salt of things dissolved by 'the acid of the spirit of nitre' (nitric acid), Baron remarks: 'The partisans of the system of affinities reply that that is due to the fact that the spirit of salt' (hydrochloric acid) 'has a closer relationship (*rapport*) with the mercury, than has the spirit of nitre.'

However, Pott had shown that *aqua fortis* poured on to corrosive sublimate (mercuric chloride) disengaged fuming spirit of salt (hydrogen chloride), which would imply the opposite affinity.

That is why I think that all these affinities and relationships which are so much vaunted have nothing absolute or real about them, and that they are merely relative to particular circumstances, and to the present state of the substances which we combine together.[85]

In picking out a reaction in which the insolubility of one of the possible products, rather than the relative affinity of the constituents, is what determines the result, Baron has of course hit on one of the practical difficulties of establishing a constant order of affinity.

Another radical critic of the theory of affinities was A. G. Monnet (1734–1817), who was an inspector-general of mines and looked at chemistry from the metallurgical point of view. His *Traité de la dissolution des métaux* consists mainly of descriptions of which substances dissolve each metal and how they do it, material which was the basis of large parts of the affinity tables produced by other chemists. However, in referring to a paper produced by Geoffroy the younger he says that 'substances act on each other according to the state in which they happen to be, rather than according to their respective affinities', and that 'the system of affinities is a beautiful chimera, better fitted to amuse our scholastic chemists than to advance that science'.[86] Similarly, in his preface he praises Kunckel, but says that it is a pity that he went wrong in 'running after the most chimerical of theories', and that in that respect he was followed by later authors.[87] He also remarks that in adding a table of contents to his book he has nevertheless avoided the absurdity of several contemporary authors who had consigned their pretended discoveries to a table.[88] Nevertheless, he does use the word 'affinity' in its chemical sense himself[89]—it is presumably only to the 'system' of affinities that he objects.

Demachy, whose own explanation of chemical combination has already been mentioned, and whose work on tables of affinity will be discussed in the next chapter, has a similarly sweeping condemnation of the whole notion of constant affinities. He objects to all previous affinity tables on the ground that their basis, that is the assumption that there was such a thing as a constant affinity, was wrong. Either they did not prove what they were intended to prove, or else instead of supporting such theories they were derived from them.

> What was my surprise when I became convinced that I was destroying the application made by Chymists of the most famous, the most vaunted, and the most universal theory, that of attraction ... One does not expose oneself lightheartedly to the risk of becoming known as the burner of the Temple of Ephesus; so I have hazarded putting my work before the judgement of the learned only to acquire new confirmations of the goodness of the theory which I dare to substitute for that which I have opposed.[90]

However, Demachy is not as sparing in his criticisms of theories of affinities as he suggests. For instance, he says that Stahl, believing that affinity could not reasonably exist except between substances which have some sort of resemblance, or at least a quality in common, made use of Becker's [*sic*] theory on the formation of substances. After outlining Stahl's system of chemical principles, which he says is implausible 'for nothing has even demonstrated to the senses what he puts forward', Demachy continues:

> However, for a long time nothing has been heard but this jargon in place of theory: if such and such a body combines with such and such another, it is because it has an *affinity*; it is because it resembles it in some of its *latus*; and the table of relationships, by procuring for Chymists the very great advantage of giving them not only the principal combinations, but their order, has done, against the intention of the author, this wrong to Chymistry, that with the words *affinity* and *latus* it was believed to be in a position to account for all phenomena; and that the feeblest Chymist believed himself a clever man.[91]

In passing we may note that opposition to the idea of a system of affinities was not necessarily associated with a desire to return to the chemistry of Stahl, as Venel seems to have wished.

Demachy goes on to criticize Freind's attempts to interpret chemistry through the mathematical and geometrical theory of attraction. Taking up, for instance, Freind's argument that a solute can diffuse through a solution because its molecules are so small that their attractive force is much greater than their weight, he replies that if they are heavier than water they will sink, however small. His chief criticism of Freind, however, is that a theory which cannot be confirmed or demonstrated in practice is meaningless. For example, Freind's theory does not explain why various metals should be differently affected by concentrated vitriolic (i.e. sulphuric) acid and by dilute vitriolic acid. Such calculations founded on so many assumptions are far from being sound chemistry. If the chemist sometimes uses the language of geometers, it should be to give more precision to his ideas, or more clarity to his expression of them, but never to pretend to explain the phenomena. What has been called 'la Chymie méchanique' (mechanical chemistry), says Demachy, is merely an abuse of mechanics. On the other hand, Macquer's explanation of affinities is merely a teaching aid, and is more acceptable.[92]

Demachy even criticizes Lesage, whose theory is much more like his own.[93] It was the very fact that they were not satisfied with theories of affinity, based on the laws of attraction and developed by followers of Stahl among others, he argues, which had induced the Academy of Rouen to require entrants for the competition of 1758 to find a physico-mechanical system of affinities. Lesage's first step, of which Demachy approves, had therefore been to re-examine the received account of attraction. Demachy also approves of Lesage's view that chemical combination is an example of the fact that bodies tend to occupy the smallest possible volume; but he disagrees with Lesage's conclusions. It is at this point that Demachy suddenly reveals his own mechanical explanation. Hitherto he has cautiously insisted that all speculations, mechanical explanations, and hypotheses of unobservable inherent properties of matter, such as universal attraction, must be treated with great reserve and subjected to the test of experiment. His approval for Lesage is due to his dislike for attraction because it involved action at a distance, a traditional Cartesian objection. He is with Boyle and Descartes in insisting that nothing can move unless there is something in contact with it to push it.

De Fourcy attacked the doctrine of affinity in a similar vein.[94]

> But I shall say in general that the terms 'relationships' and 'affinities' present to the mind merely words empty of sense; thus it would be a poor explanation of the combination of one substance with another for a Chymist to say that these two substances have an *analogy*, a *relationship* or an *affinity* which causes the unions; that would be explaining the obscure by the more obscure.[95]

De Fourcy is a partisan of Meyer's theory that acidic and caustic properties were due to the substances concerned's containing an acid called *acidum pingue* (fatty acid), a theory which was directly contradicted by Black's work on the alkaline earths, at least as far as caustic alkalis were concerned.

De Fourcy hence attacks Stahl's theory that one substance combined with another by means of a principle which they had in common, or *latus*.

Metals can be attacked by acids, or so it is claimed, only by their *latus* of phlogiston. I cannot understand how phlogiston facilitates the union of an acid with a metal, when it has a relation with acids itself only by its *latus* of *acidum pingue*. Phlogiston, on the contrary, is an obstacle which must be overcome to make possible the penetration of the metallic earth, and whenever a metal is dissolved by an acid, the phlogiston is always destroyed and the metal becomes completely calx [oxide].[96]

Demachy exaggerates the extent to which most chemists of his day relied on the concept of affinity as a theoretical explanation. Even those who published affinity tables or theories of affinity generally claimed at least to start from the observed facts and to construct merely a plausible hypothesis, or to find merely an apparent pattern in the phenomena, which might suggest a useful explanation, rather than to start with the assumption that some particular theory was correct and then fit the facts to it. Admittedly those who wrote accounts of possible theories or regularities commonly seemed to be fairly certain that they were correct; but that after all is natural enough, and in any case it was not often the producers of the most fruitful results who wrote in favour of particular theories. Between the two extremes of condemning entirely even the use of such words as 'attraction' or 'affinity', and of accepting unverifiable hypotheses about mechanical explanations of attraction or affinity, most held the middle course of using the words to mean simply the observed tendency of chemical substances to combine with each other but emptying them of any theoretical implications.

'Attraction' and 'affinity' come to have the same meaning

It is sometimes said that the theories of affinity and attraction were rival theories, or that one replaced the other. In their origins as physical explanations that was true. Nevertheless, among chemists the two words came to mean the same thing, although it was remembered that their origins were different. For some time the word 'affinity', with its connotations of a cousinly relationship or at least a similarity, was more common on the Continent of Europe, and the Newtonian word 'attraction', with its connotation of action at a distance, was more used in Britain. However, by the 1770s, though it was not forgotten that the words had once had these connotations, they had become in normal usage interchangeable with each other and with the German equivalent of 'affinity', 'verwandtschaft'. Professor Guerlac, in the brilliant paper which he read at the celebrations of Dalton's bicentenary, argued that on the contrary the word 'attraction' had implications which 'affinity' had not. He suggested that 'men were led, as a result, to think of interparticulate *forces*, and therefore of entities that, in principle, could be measured', and that the notion that the forces were between particles or atoms which had unchanging properties was a particularly important implication of Newtonianism.[97]

Leading men to think in the way described was certainly an important effect, but it was an effect of conceiving of affinity as a constant property of particles as well as of accepting the concept of attraction. It is no doubt true that in the early eighteenth century such concepts were associated with Newtonian ideas, held mostly by physicists, and not altogether shared by those, mostly chemists, who spoke of 'affinity'. Nevertheless, attraction and affinity were not much different in their practical implications for chemistry even in the early years of the eighteenth century, and by the 1760s or so the Newtonian approach had been generally adopted even in France. The urge to quantify, and the belief that chemistry was a matter of the properties of particles were widespread, and attached to the word 'affinity' as much as to the word 'attraction', both having lost any connotations of causal explanations.

Although chemists in the eighteenth century virtually all believed that matter consisted of particles, and although some of the implications of that belief were helpful to them, speculations about the shapes and sizes and mechanisms of particles could not be tested experimentally and did not contribute much if at all to solving the problems that were immediately significant in chemistry. In other words, the difference between attraction and affinity was important to physicists but not to chemists. That is shown by the way in which the words were used. In defining whichever of the two they chose to use, chemists did sometimes show awareness of the original difference in meaning; but when it comes to using them the difference disappears.

Joseph Black (1728–99), a stickler for precision, summed up the differences between their strict senses in his lectures. He says that chemists abroad at first

> objected to the word attraction, as implying either an active quality in matter, which we cannot conceive to be possessed of activity, or some connecting intermediate substance, by which the particles of bodies were drawn together, and for the existence of which no proof is offered. They therefore chose to substitute the term AFFINITY instead of ATTRACTION.

Black considered this unreasonable, since 'attraction' brings out the similarity between chemical, gravitational, electric, and magnetic attraction, and

> affinity implies, or suggests, some similarity which, in most cases, is not agreeable to fact, seeing that we generally observe the greatest dissimilarity in those bodies which are eminently prone to unite.[98]

Whichever word is used, the definition given by chemists was usually the same. It is an observable fact that substances do tend to unite, and 'affinity', 'attraction', or 'verwandtschaft' is the name given to this tendency, whatever the cause may be. Thus the name can be used by chemists to describe the phenomenon, which is of course extremely important to them, and is taken to refer to some intrinsic property of the particles of matter without any speculation or implication about the cause of the property. Such definitions are derived from one level of interpretation of Newton's work, for as has already been pointed out he cautiously wrote, in spite of his theories:

How these Attractions may be performed, I do not here consider. What I call Attraction may be performed by impulse, or by some other means unknown to me. I use that Word to signify in general any Force by which Bodies tend towards one another, whatsoever be the cause.[99]

Similarly Desaguliers, writing as a physicist and a follower of Newton, wrote:

When we use the Words Gravity, Gravitation or Attraction; we have a regard not to the Cause, but to the Effect; namely to that Force, which Bodies have when they are carried towards each other, which (at equal distances) is always proportionable to their Quantity of Matter; whether it be occasioned by the Impulsion of any subtile Fluid, or by any unknown and unmechanical Power concomitant to all Matter.[100]

Chemists do not usually even assume, at least until the fourth quarter of the century, that anything which might properly be called a 'force' is at work. For instance Black's teacher and predecessor William Cullen (1710–90) said in his lectures:

The Combination of different Particles depends solely upon Attraction, a term first adopted by Sir Isaac Newton to signify the power by which Bodies move towards one another without determining whether they are really drawn or forced towards one another.[101]

As has already been mentioned, J. R. R. Christie has shown that Cullen used a theory, probably derived from Bryan Robinson, of an ether that was made up of particles that repelled each other. The theory developed and changed between the late 1740s and the early 1760s. At one time he argued that chemical attraction depended on the relative strength of the attraction of cohesion between the particles of ordinary matter and the repulsion between the particles of the ether. Later he explained attraction through differences in the density of the ether outside and inside a body. Cullen was particularly concerned with the evolution of heat or cooling which occurred during chemical change, and tried to relate the selectiveness of chemical combination to such thermal changes. Christie relates

Cullen's belief in an ether to the ideas of his contemporaries David Hume and Adam Smith, and the notion of the ether, rather than God, as a final cause to secularization. Cullen's assertion that chemistry was a proper part of philosophy is linked with the usefulness of chemical theory in the growing economy of Scotland. Cullen's pupil Joseph Black, whose best known work was on specific and latent heat, may well have believed that heat was a substance like ether, though he was cautious about revealing such speculations.

Already in 1763–5 at least one British chemist, William Lewis (1708–81), who published the first affinity table in English, used the words 'attraction' and 'affinity' as equivalents. He wrote in his *Commercium philosophico-technicum*:

> To the grand active power, called *attraction*, in the mechanical philosophy, what corresponds in the chemical is generally distinguished by another name, *affinity*.
>
> The mechanical attraction obtains between bodies considered as one whole, and between bodies of the same, as well as of different kinds. It obtains while the bodies are at sensible distances; and the comparative forces, with which they tend together at different distances, are objects of calculation. When the attracting bodies have come into the closest contact we can conceive, they still continue two distinct bodies, cohering only superficially, and separable by a determinate mechanical force.
>
> The chemical attraction, or affinity, obtains between bodies as being composed of parts, and as being of different species of matter from one another. It never takes place while the two bodies are at any sensible distance; and when they are brought into the closest contact, there is frequently necessary some other power, as fire, to excite their action upon one another. In proportion as this action happens, they are no longer two bodies, but one; the affinity consisting in the intimate coalition of the parts of one body with those of another ...'[102]

Lewis's distinction between the mechanical philosophy and the chemical philosophy is significant. The implication that 'attraction' and 'affinity' have the same meaning, but that one is used in physics and the other in chemistry, is found elsewhere, for instance in Tessari's *Chymiae elementa in aphorismos digesta* of 1771: 'All substances, especially the simpler ones, which are usually examined by the manifold toil of Chemistry, show in turn a certain particular tendency which the Physicists call *attraction*, the chymists *affinity*.'[103]

At the end of the century, A. F. de Fourcroy (1755–1809) accepted that the two words meant the same thing, but recorded that 'attraction' had replaced 'affinity', because of the connotations of the latter word. By that time the original French objections to Newtonianism, and the consequent preference for the term 'affinity', had been overcome and forgotten. Thus Fourcroy wrote in 1800: 'The Affinity of aggregation of the chemists is nothing more than the attraction which exists between particles of like nature adhering to each other.'

A few pages further on he continued:

The attraction of composition, which was formerly called chemical affinity, because it was supposed to be produced by a natural relation between the bodies in which it is observed, differs from the attraction of aggregation in the circumstance, that this last never acts but between similar particles ...[104]

A few years earlier, however, he had not been troubled by this objection to the use of the word 'affinité', which had normally been preferred in France, and had called it 'this wonderful force, established between all natural bodies, by which they attract each other reciprocally'.[105] Similarly, in the memoir of 1784 in which he tried to quantify the force of affinity, he seemed to find no objection to the use of the word.[106] Thus the urge to quantify is not necessarily connected with the Newtonian term. Also, Fourcroy, who was active in politics before and after the Revolution, preferred the word 'affinity' before the Revolution and the dissolution of the old *Académie*, and 'attraction' afterwards. It is hard to see any political significance in the terminology.

In Germany, F. A. C. Gren (1760–98) made a straightforward identification between the two concepts. He wrote:

In chemistry, the action of that power, which is naturally inherent in heterogeneous substances, and by virtue of which they unite together in various degrees of intensity, goes by the name of *chemical affinity*, or *elective attraction*; and that substance is supposed to possess a *nearer* or greater affinity with another, which is more strongly attracted by it, than to a third which is less.[107]

A typical chemist's definition of affinity is given by Baumé: 'We define chemical affinities as a tendency which the parts of matter have to unite and adhere together, whether these parts be homogeneous or heterogeneous.'[108]

Guyton de Morveau wrote of affinity:

This is the name given in Chemistry to the force with which substances of a different nature tend to unite. This term which in its proper and original sense indicates only a close bond of relationship, which, in figurative language, hardly applies except to moral or metaphysical connections, is today the expression of a purely physical action.[109]

Thus Guyton, writing in the 1780s, was prepared to call 'affinity' a force, as was Fourcroy; and both chemists, as we shall see, were interested in measuring its magnitude.

However Guyton does make a distinction between 'attraction', which he uses to mean gravitational attraction, and 'affinity', by which he means what some others called 'chemical attraction'.

I am far from calling into doubt that what Chemists have named affinity is an effect which proceeds immediately from the property which bodies have of attracting each other reciprocally; but there is always *attraction* between all matters, and we shall soon see that there is not always affinity; we therefore need a particular sign to specify that intensity of attractive power on which chemical phenomena depend; it is enough to reserve for it the name affinity, the allocation of which is consecrated by a long possession, and which expresses the same thing in fewer words, than does elective attraction, which we shall be able to employ as a synonym. The former will still have the advantage of being less meaningful for Chemists, if there are any, to whom the identity of attraction and affinity does not yet seem quite demonstrated.[110]

Bergman's influential *Dissertation on elective attractions* begins with a similarly empirical definition.

It is found by experience, that all substances in nature, when left to themselves, and placed at proper distances have a mutual tendency to come into contact with one another. This tendency has long been distinguished by the name of *attraction*. I do not propose in this place to enquire into the cause of these phaenomena; but, in order that we may consider it as a determinate power, it will be useful to know the laws to which it is subject in its operations, though the mode of agency be yet unknown.[111]

Bergman gives his reason for preferring the term 'attraction' a little further on:

What I here call attraction, others denominate affinity: I shall employ both terms promiscuously in the sequel, though the latter, being more metaphorical, would seem less proper in philosophy.[112]

By 1775 chemists were less worried by the sort of objection to even the word 'attraction', not to speak of the concept, that Demachy had expressed so strongly. Bergman, however, like Baumé, is prepared to go further, and identify as a matter of definition the force producing chemical attraction with that responsible for gravitation. He also speculates a little, in the manner of Buffon, on the effect of the size, shape, and relative position of the particles of matter on the force of attraction between them, as well as the effect of the earth's attraction. He concludes that these factors will complicate matters so that the inverse square law of attraction cannot be applied to what he calls 'contiguous attraction' as if the particles were points, as it can be to the attraction between heavenly bodies which are at vast distances apart; but neither can the shape and position of the particles be determined. The attraction of bodies must therefore be determined experimentally in each case.[113]

John Berkenhout, a close follower of Bergman, also made clear his reasons for preferring to speak of 'attraction'; but, unlike Bergman, he distinguishes between gravitation and chemical attraction. He wrote:

All bodies which are the object of Chemistry, discover a propensity to unite with other bodies; they are drawn together by mutual attraction; but this attraction differs from gravitation in not acting proportionably to the quantity of matter.

This property, from a supposed similitude in the principles of certain bodies, hath been called *affinity*, but improperly, for many bodies which unite most eagerly are totally dissimilar in their nature and properties.[114]

Although he dislikes the word 'affinity', Berkenhout does not question that the property to which some people give that name is the same one that he prefers to call 'attraction'.

These examples show that although a number of eighteenth-century chemists were interested in the cause of chemical attraction or affinity, most of them either kept their speculations about the cause separate from what they wrote about actual chemistry or did not speculate about it at all; and that the definition which they gave of chemical attraction or affinity was simply the observable tendency of substances to combine. They differed indeed on the relationship between chemical attraction and other kinds of attraction, and some had a preference for either the word 'attraction' or the word 'affinity' because of the connotations of their origins. Nevertheless, the operational definition was the same in either case. If we examine the use that chemists of the 1760s and 1770s made of the words 'affinity' and 'attraction' in describing and interpreting experimental results, we find that they do indeed empty those words of theoretical implications and use them simply to mean 'tendency to combine'.

For instance, Henry Cavendish (1731–1810) in the second of his 'Three papers containing experiments on factitious air' speaks of the fixed air (carbon dioxide) discharged from marble as having 'a greater affinity to fixed alcali' (sodium or potassium hydroxide) than the rest. As the fixed air is of the same chemical nature in each case, there is clearly no implication of greater similarity.[115] The word 'affinity' is simply the normal colourless word for 'tendency to combine'. Yet Cavendish clearly had no reason to object to the word 'attraction' in other contexts. Indeed, Dr McCormmach has argued that Cavendish's multifarious researches are all related to a single basic aim, that of discovering the forces of particles, which like Bosocovich he preferred to think of as point-centres of forces.[116] Cavendish was, of course, rather more in touch with the upper crust of natural philosophers, or physicists as they would now be called, than most practical chemists; and in fact after the refutation of the phlogiston theory he moved on to work on electricity.

The theologian, political pamphleteer, and polymath Joseph Priestley (1733–1804), who had also worked on electricity, wrote in his *Experiments and observations on different kinds of air*: 'Inflammable substances can only be those which, in a certain degree of heat, have a less affinity with the phlogiston they contain, than the air, or some contiguous substance, had with it.'[117]

Clearly Priestley is suggesting that affinity varies with temperature, and there is no trace of the original meaning of the word. Later in the same work he speaks of 'a real acid air, having an affinity with water, similar to other acid airs'.[118] Plainly in saying that an acid air (i.e. a gas with acidic properties) has an affinity with water he does not in the least intend to suggest that it is similar in nature to water. He means merely that it has a tendency to combine with water.

However, in a private letter Priestley was quite ready to speculate about the relationship between attraction and repulsion, though there is no suggestion of any opinion about the cause of them. In writing to Joseph Wedgwood about his claim to have converted water to permanent air by strong heating (a part of his efforts to refute the anti-phlogistic theory), Priestley describes how he had produced air by combining water with quicklime (calcium oxide) and heating it strongly. He believed that in this way he would be able to expose the water to strong heat without its evaporating, as it would be held in 'firm combination'. He adds, possibly with Boscovich's theory in mind: 'I had also a general idea that if the parts of the bodies be rarefied beyond the sphere of *attraction* they will be in a sphere of repulsion to each other.'[119]

Professor Schofield, indeed, has argued that Priestley, who like Cavendish was well acquainted with the ideas of attraction and repulsion from his work on electricity, was

> a natural philosopher in that English mechanistic tradition, including the work of Boyle, Newton, John Keill, John Freind and Stephen Hales, for which all natural phenomena must ultimately be explained by the primitive particles, their actions, and the forces between them. Priestley's fondness for Boscovichian atomism is consistent with that tradition.[120]

It is true that in his psychology and metaphysics, and even in his theology, Priestley was a reductionist, believing that everything could be explained materially and mechanically. Presumably he believed that the ultimate explanations of chemical phenomena were of the same kind. Nevertheless, that belief hardly appears in his writings on chemistry. There he is concerned with reporting the results of his experiments and using them as evidence for the composition of substances; but he scarcely refers to mechanical explanations. At any rate, it seems clear that when he speaks of 'affinity' he does not mean anything different from what he means when he uses the word 'attraction' in a chemical context, that is 'tendency to combine'.

Many examples from the writings of other chemists might be quoted, but one more will suffice. C. W. Scheele (1742–86) wrote a paper of which the title may be translated as 'Some incidental remarks on the affinity of bodies' in which he refers to 'attraction' and says, 'This earth of alum' (aluminium oxide, presumably with traces of sulphate) 'can scarce be without vitriolic acid;' (sulphuric acid) 'for we know, that a part of this acid is very difficult to be separated from it; and, on this account, it could not attract any fixed air.'[121]

Plainly the words in the original represented by 'affinity' and 'attraction' refer to the same concept, that is the tendency of substances to combine with each other, and do not imply any particular mechanism.

The position is summed up by William Nicolson (1753–1815) in the introduction to his edition of Kirwan's *Essay on phlogiston*, when he says that various bodies 'combine and exhibit those various dispositions to unite or separate, which have been referred to specific attractions, called the chemical affinities'.[122] In other words, to Nicholson in 1789 the words 'affinity' and 'attraction' in chemistry mean exactly the same thing, that is a disposition to combine, and no speculation about the cause of it can be read into either.

Notes

1. Gowin Knight, *An attempt to demonstrate that all the phaenomena in nature may be explained by two simple active principles, attraction and repulsion ...*, London, 1748; 2nd edn, London, 1754, p. 1.
2. G. L. Lesage, *Essai de chymie méchanique*, Geneva, 1762; MS introduction in the printed copy deposited at the Bibliothèque Publique et Universitaire, Geneva.
3. P.-J. Macquer, *A dictionary of chemistry*, (trans. by James Keir from *Dictionnaire de chymie*, 2 vols, Paris, 1766), 2 vols, London, 1771, Vol. 1, P. 324.
4. T. O. Bergman, *Dissertation on elective attractions* (English trans., probably by T. Beddoes, of the version in Vol. III of Bergman's *Opuscula physica et chemica*, 1783, of 'Disquisitio de attractionibus electivis', *Nova Acta regiae Societatis Scientiarum Upsaliensis*, **2**, 1775, pp. 161–250), London, 1785; repr. with introduction and appendices by A. M. Duncan, London, 1970, p. 4.
5. A.-L. de Lavoisier, 'Mémoire sur l'affinité du principe oxygine', *Mémoires de l'Académie Royale des Sciences*, 1782, p. 530.
6. See Chapter 1, p. 16; R. E. Schofield, *Mechanism and materialism. British natural philosophy in an age of reason*, Princeton, 1970; and A. Thackray, *Atoms and powers*, Cambridge, Mass., 1970.
7. See C. C. Gillispie, 'The *Encyclopédie* and the Jacobin philosophy of science: a study in ideas and consequences'; L. Pearce Williams, 'The politics of science in the French Revolution'; commentaries by H. B. Hill and H. Guerlac on these two papers in Marshall Clagett (ed.), *Critical problems in the history of science*, Madison, 1959, pp. 255–320; and W. L. Scott, 'The impact of the French Revolution on English science', *Mélanges Alexandre Koyré*, Vol. II, ed. F. Braudel, Paris, 1964, pp. 475–95, esp. pp. 482–3.
8. J. Quincy, *Pharmacopée universelle raisonnée*, trans. Clausier, Paris, 1749, p. 8.
9. Ibid., pp. 55 ff.
10. Jean Philippe de Limbourg, *Dissertation sur les affinités chymiques*, Liège, 1761, p. 9.
11. Ibid., pp. 12–14.
12. Ibid., p. 30.
13. Ibid., p. 45.

14. See W. A. Smeaton, 'Guyton de Morveau and chemical affinity', *Ambix*, **11**, 1963, pp. 55–64. Guyton's views on affinity first appeared in an essay in his first book, *Digressions académiques*, Dijon, 1772, pp. 271–377; but similar theories also appear in the articles on *Affinité*, *Crystallisation*, *Dissolution*, and *Equipondérance*, which he wrote for D. Diderot, *Supplément à l'Encyclopédie*, 4 vols, Amsterdam, 1776–7.
15. Cf. Marie Boas, 'The establishment of the mechanical philosophy', *Osiris*, **10**, 1952, pp. 519–20. It should be remembered that not much information about Newton's ether theory was available in print until 1744.
16. John Keill, *An introduction to natural philosophy or lectures read in the University of Oxford, Anno Dom. 1700*, 4th edn., London, 1745 (1720), pp. 46–50. See also Anita Guerrini and Jole R. Shackelford, 'John Keill's *De operatione chymicarum ratione mechanica*', *Ambix*, **36**, 1989, pp. 138–52.
17. Keill, *An introduction to natural philosophy*, 4th edn, London, 1745, pp. iii-iv.
18. Ibid., pp. 94–5.
19. John Freind, *Chymical lectures in which almost all the operations of chymistry are reduced to their true principles and the law of nature. Read in the Museum at Oxford*, 1704, 2nd edn., London, 1729 (1712), pp. 7–10.
20. Ibid., pp. 95–104.
21. Ibid., pp. 146–7.
22. Keill, *Philosophical Transactions of the Royal Society*, 1708, pp. 97–110.
23. Stephen Hales, *Vegetable staticks*, London, 1727; Oldbourne reprint, London, 1961, p. 139.
24. Ibid., pp. 118–19 and 121.
25. Henry Pemberton, *A view of Sir Isaac Newton's philosophy*, London, 1728; Roger Cotes, *Hydrostatical and pneumatical Lectures*, London, 1738.
26. W. J. van 'sGravesande, *Mathematical elements of natural philosophy*, trans. J. T. Desaguliers, London, 1720.
27. J. T. Desaguliers, *A course of experimental philosophy*, 2 vols, London, 1734, pp. 16–17. Cf. P. van Musschenbroek, *Epitome elementorum physico-mathematicorum in usus academicos*, Leiden, 1726; trans. into English by J. Colson, 2 vols, London, 1774.
28. J. Marzucchi, *Chymiae elementa*, Padua, 1751.
29. See Schofield, *Mechanism and materialism*, Princeton, 1970, and Thackray, *Atoms and powers*, Cambridge, Mass., 1970.
30. Anon., *Nouveau cours de chymie suivant les principes de Newton & de Sthall*, 2 vols, Paris, 1723; 2nd edn, Paris, 1737, Vol. I, p. 146.
31. Ibid., pp. 74–6.
32. Ibid., p. 76.
33. Ibid., pp. 77–8.
34. Ibid., pp. 79–118.
35. Ibid., pp. 152 ff. Cf. Newton, *Correspondence*, ed. H. W. Turnbull, Vol. I, Cambridge, 1959, pp. 365–6.
36. Anon., *Nouveau cours de chymie*, 2nd edn, Paris 1737, Vol. I, pp. 152 ff.
37. Macquer, writing to Bergman, *T. Bergman's foreign correspondence*, Stockholm, 1965, p. 230; Fourcroy, in *Encyclopédie méthodique*, Vol. iii, Paris, 1796, p. 740; D. C. R. Weigel, *Grundriss der reinen und angewandten Chemie*, 2 vols, Greifwald, 1777, Vol. I, p. 11; J. F. Gmelin, *Geschichte der Chemie*, 3 vols, Göttingen, 1797–9, Vol. II, p. 682; H. Kopp, *Geschichte der Chemie*, 4 vols, Brunswick, 1843–7, Vol. IV, p. 179. See J. R. Partington, *History of chemistry*, Vol. III, London, 1962, p. 58.

38. In Lemery, *Cours de chymie*, ed. Baron, Paris, 1756, p. iii.
39. E.-F. Geoffroy, *Mémoires de l'Académie Royale des Sciences*, 1718, pp. 202–12.
40. A. Thackray, 'Quantified chemistry—the Newtonian dream', in D. S. L. Cardwell (ed.), *John Dalton and the progress of science*, Manchester, 1968, p. 98.
41. E.-F. Geoffroy, trans. G. Douglas, *Treatise of the fossil, vegetable and animal substances that are made use of in Physick*, London, 1736. See also W. A. Smeaton, 'E. F. Geoffroy was not a Newtonian Chemist', *Ambix*, **18**, 1971, pp. 212–14.
42. *Catalogus librorum ... Stephani-Francisci Geoffroy*, Paris, 1731. No. 1502, 'Traité des reflexions, refractions et inflexions des couleurs de la lumiére, Lond. 1704 in 4 (en anglois)', is presumably Newton's *Opticks*, Geoffroy's copy of which is now in the library of Cornell University (see H. Guerlac, 'Newton in France—two minor episodes', *Isis*, **53**, 1962, pp. 219–21).
43. I. B. Cohen, 'Isaac Newton, Hans Sloane, and the Académie Royale des Sciences', *Mélanges Alexandre Koyré*, Vol. I, Paris, 1964, pp. 61–116; H. Guerlac, 'The background to Dalton's atomic theory', in Cardwell (ed.), *John Dalton and the progress of science*, Manchester 1968, p. 73.
44. Venel, article 'Chymie' in the *Encyclopédie*, Vol. III, Paris, 1753, repr. by Fourcroy in *Encyclopédie méthodique*, Vol. III, Paris, 1796, p. 302; quoted by Partington, *History of chemistry*, Vol III, London, 1962, p. 58.
45. I am indebted to Miss J. P. S. Ferguson, Librarian of the Royal College of Physicians, for allowing me to use the manuscript of Plummer's lectures, which are entitled *Opera chemica*.
46. A. Plummer, 'Experiments on neutral salts', *Essays and observations physical and literary*, 2nd edn, Edinburgh, 1771, pp. 374–6.
47. Roger Boscovich, *Philosophiae naturalis theoria reducta ad unicam legem virium in natura existentium*, Vienna, 1758, pp. 37 ff; much improved edn, Venice, 1763; English trans. by J. M. Child, *A theory of natural philosophy*, 2nd edn, Cambridge, Mass., 1966. See also L. L. Whyte (ed.), *Roger Joseph Boscovich*, London, 1961, and the proceedings, published by the Istituto della Enciclopedia Italiana, of the conference on the work of Boscovich held at the Istituto in May 1988.
48. Boscovich, ibid., pp. 84–279, esp. pp. 96–125.
49. Newton, *Opticks*, New York, 1952 (1704), p. 395.
50. Gowin Knight, *An attempt ...*, 2nd edn, London, 1754, p. 10 (see Note 1).
51. Loc. cit.
52. Ibid., pp. 19 ff.
53. Ibid., p. 23.
54. Ibid., pp. 28 ff.
55. Loc. cit.
56. J. Michell, *Treatise of artificial magnets*, 2nd edn, Cambridge, 1751 (1750), pp. 17–19; quoted by A. Woolf, *A history of science, technology and philosophy in the eighteenth century*, 2nd edn, revised by D. McKie, London, 1952, pp. 269–70.
57. Plummer, 'Experiments on neutral salts', pp. 374–6 (see Note 46).
58. A. Baumé, *Chymie expérimentale et raisonnée*, 3 vols, Paris, 1773, Vol. I, p. 27.
59. Macquer, *A dictionary of chemistry*, 2 vols, London, 1771, Vol. II, p. 195.
60. L. B. Guyton de Morveau, *Encyclopédie méthodique*, Paris and Liège, 1789, pp. 540–1.
61. Baumé, *Chymie experimentale et raisonée*, 3 vols, Paris, 1773, Vol. I, p. 27.
62. Baumé, *Avant-Coureur*, 16 March, 1772, p. 728, and Rozier's *Observations*,

1773, 1, p. 8; reprinted in R. Davy, *L'Apothicaire Antoine Baumé*, Cahors, 1955, pp. 121–2.
63. Lavoisier, Rozier's *Observations*, 1773, 1, p. 10.
64. See Partington, *History of chemistry*, Vol. III, London, 1962, pp. 749–54, and J. R. Partington and T. S. Wheeler, The life and work of William Higgins, chemist, Oxford and London, 1960.
65. Bryan Higgins, *A philosophical essay concerning light*, Vol. I (only one volume issued), London, 1776, pp. xix-xxi. See also J. R. Partington, 'Origins of the atomic theory', *Annals of Science*, **4**, 1939, pp. 245–82.
66. Bryan Higgins, ibid., p. xxxviii.
67. Ibid., p. xxxix.
68. Ibid., p. xxxvi.
69. Ibid., pp. xlvii-xlviii.
70. Ibid., p. xliv.
71. Ibid., e.g. p. xxii for the earthy element.
72. Ibid., pp. 124–5.
73. Ibid., pp. 44–61.
74. Ibid., pp. 120–3.
75. Ibid., pp. 21–3.
76. Ibid., pp. xxxviii-xxxix and 130–45.
77. Jean-Louis Leclerc, Comte de Buffon, *Histoire naturelle*, 'Seconde vue de la nature', Paris, 1765, pp. xii-xvi. Buffon had previously been involved in a controversy over the law of attraction with Clairaut, who thought he had detected a faster movement of the moon's apogee than that required by Newton's inverse square law of gravitation, and suggested that as well as the inverse square term the law should include a second term inversely proportional to the fourth power of the distance, which would have appreciable effects at very short distances. Buffon argued that the law governing gravitation could not have more than one term, and Clairaut eventually recalculated the movement of the moon's apogee and admitted that Buffon was right. (Alexis Claude Clairaut, *Mémoires de l'Académie Royale des Sciences*, 1745 (read 1747–9), pp. 329–64, 529–48, 577–9, 583–7; Buffon, *Mémoires de l'Académie Royale des Sciences*, 1745, pp. 493–500, 551–2, 580–3.)
78. Guyton de Morveau, *Encyclopédie methodique*, Paris and Liège, 1789, pp. 540–1.
79. Torbern Bergman, *Dissertation on elective attractions*, London, 1785 (repr. 1970), pp. 2–3.
80. G. L. Lesage, *Essai de chimie méchanique*, Geneva, 1762, pp. 14–30 and *passim*.
81. Ibid., p. 52.
82. J. F. Demachy, *Recueil de dissertations*, Amsterdam, 1774, pp. 150–2, 211–26, 235.
83. See A. M. Duncan, 'Particles and eighteenth-century concepts of chemical combination', *British Journal for the History of Science*, **21**, 1988, pp. 447–53.
84. N. Lemery, *Cours de chymie*, ed. T. Baron d'Hénouville, Paris, 1756, p. 79 note.
85. Ibid., p. 233 note.
86. A. G. Monnet, *Traité de la dissolution des métaux*, Amsterdam and Paris, 1775, p. 55.
87. Ibid., pp. iii-iv.
88. Ibid., p. vii.
89. Ibid., p. 159.

90. Demachy, *Recueil de dissertations*, Amsterdam, 1774, p. 83.
91. Ibid., p. 103.
92. Ibid., pp. 107–10.
93. Ibid., pp. 137–50.
94. De Fourcy, 'Observation sur le tableau du produit des affinités chymiques', Rozier's *Observations*, 1773, 2, pp. 197–204. Mentioned by E. W. J. Neave, *Annals of Science*, **8**, 1952, p. 28.
95. De Fourcy, Ibid., p. 197.
96. Ibid., pp. 198–9.
97. Guerlac, 'The background to Dalton's atomic theory' (see Note 43), p. 83.
98. J. Black, *Lectures on the elements of chemistry*, ed. J. Robinson, 2 vols, Edinburgh, 1803, Vol. I, pp. 266–7.
99. Newton, *Opticks*, New York, 1952 (1704), p. 375.
100. Desaguliers, *A course of experimental philosophy*, 2 vols, London, 1734, Vol. I, p. 6.
101. MS of Cullen's lectures, as delivered about 1757–8, in the possession of Dr W. A. Smeaton, to whom I am indebted for allowing me to use the MS. See further J. R. R. Christie, 'Ether and the science of chemistry: 1740–1790', in G. N. Cantor and M. J. S. Hodge (eds), *Conceptions of ether: studies in the history of ether theories 1740–1900*, Cambridge, 1981, pp. 85–110.
102. W. Lewis, *Commercium philosophico-technicum*, London, 1763, pp. iv–v. Quoted by Nathan Sivin, 'William Lewis (1708–81) as a chemist', *Chymia*, **8**, 1962, p. 75.
103. L. Tessari, *Chymiae elementa in aphorismos digesta*, Venice, 1772, p. 8: 'Corpora omnia simpliciora praesertim, quae multiplici Chymico labore scrutari solent, certam quamdam tendentiam invicem ostendunt quam Physici *attractionem*, chymici *affinitatem* appellant.'
104. A. F. de Fourcroy, *A general system of chemical knowledge* (trans. by William Nicholson of *Système de connaissances chimiques* of 1800), 11 vols, London, 1804, Vol. I, p. 90 and p. 95.
105. Fourcroy, *Elémens d'histoire naturelle et de chimie*, 4th edn, 5 vols, Paris, 1791 (1782), Vol. I, p. 45.
106. Fourcroy, 'Exposé d'une nouvelle manière d'expliquer, à l'aide des nombres, la cause des décompositions opérées par les affinités doubles', *Mémoires et observations de chimie*, Paris, 1784, pp. 308–23.
107. F. A. C. Gren, *Principles of modern chemistry* (trans. by Gruber of *Grundriss der Chemie* of 1796–7), 2 vols, London, 1800, Vol. I, p. 45. The word in the German text is *verwandtschaft*, of which 'affinity' is a precise equivalent.
108. Baumé, *Chymie experimentale et raisonée*, 3 vols, Paris, 1773, Vol. I, pp. 19–20.
109. Guyton de Morveau, *Encyclopédie méthodique*, Paris and Liège, 1789, p. 535.
110. Ibid., p. 536.
111. Berman, *Dissertation on elective attractions*, London, 1785 (repr. 1970), pp. 1–2.
112. Ibid., p. 7.
113. Ibid., pp. 2–3.
114. J. Berkenhout, *First lines of the theory and practice of philosophical chemistry*, London, 1788, p. 134.
115. H. Cavendish, 'Three papers containing experiments on factitious air', *Philosophical Transactions of the Royal Society*, 1766, pp. 141–83; repr. in Cavendish, *Scientific papers*, London, 1921, Vol. II, p. 77. Quoted by A. J. Berry, *Henry*

Cavendish, London, 1960, p. 50.

116. See Russell McCormmach, 'Henry Cavendish: a study of rational empiricism in eighteenth-century natural philosophy', *Isis*, **60**, 1969–70, pp. 293–306.
117. Joseph Priestley, *Experiments and observations on different kinds of air*, 2nd edn, 3 vols, London, 1775–7 (1774–7), Vol. I, p. 141.
118. Ibid., Vol. II, p. 168.
119. R. E. Schofield, *A scientific autobiography of Joseph Priestley*, Cambridge, Mass., 1966, p. 215.
120. Ibid., p. 195; id., 'Joseph Priestley, the theory of oxidation and the nature of matter', *Journal of the History of Ideas*, **25**, 1964, pp. 285–94.
121. C. W. Scheele, *Chemical essays* (trans. T. Beddoes), London, 1786, p. 353.
122. Kirwan, *Essay on phlogiston*, 2nd edn, London, 1789 (1784), pp. v-vi.

4
Tables of affinity and of chemical attraction

Geoffroy

In this chapter we shall first describe and discuss the evolution of tables of affinity and chemical attraction and some criticisms of them in the eighteenth century, and then show how they illustrate the development of the autonomy of chemistry as a branch of natural philosophy with its own appropriate concepts distinct from those of physics.

It was noticed very early in the history of chemistry that a given substance tends to combine with some substances more readily than with others, and the idea of making out a list of substances in order of their readiness to combine with other particular substances or kinds of substances was not new in the early eighteenth century. For instance, such an order of reactivity is referred to, as has been mentioned earlier, by pseudo-Geber, by Paracelsus, and in a later period by Newton, by Glauber, and by Stahl.[1] The credit for devising what came to be known as an affinity table, however, belongs entirely to Étienne-François Geoffroy (1672–1731).[2] Son of a wealthy Parisian apothecary, Geoffroy was intended by his father to follow him as an apothecary, while the younger brother Claude Joseph (1685–1752) became a physician. Étienne-François studied botany, anatomy, and chemistry (in which his interest was stirred by Homberg, a friend of his father's) in Paris, and then was sent to Montpellier to study pharmacy. However, he admitted to his father that he really wanted to be a physician, and was allowed to exchange professions with his brother.

It was in 1698 that he made the visit to London which has been mentioned in the previous chapter, as physician to Count Taillard, who was acting as Ambassador Extraordinary. Geoffroy was elected a Fellow of the Royal Society (not, of course, quite the accolade for distinguished research that it is now) after only six months in London. He then visited Holland and Italy with the Abbé Louvois. As we have seen, he became the link between the Royal Society and the Académie Royale in Paris after his election to the latter in 1699. He became Professor of Chemistry at the Jardin du Roi, a famous centre of public teaching in science, in 1712, after deputizing for the previous professor, Fagon, for five years.

He had also become Professor of medicine at the Collège Royale in 1709. In 1726 he was elected Dean of the Faculty of Medicine of Paris, though his opponents claimed that the election procedure had been irregular; and he was elected unopposed for a further two-year term in 1728. He is said to have shown particular skill in handling as Dean a dispute between the physicians and the surgeons, who of course still formed a separate profession at that time. It is also said of Geoffroy that he was modest and as attentive to his poor patients as to the rich ones.

At any rate it is clear that he was a respected and orthodox chemist of his time. His theoretical assumptions may be inferred from his published lectures. He argued that there were only three true elements, Fire being an active principle and Water and Earth passive principles. Salt arose from the combination of these three, and Sulphur or Oil from the combination of the three principles with Salt. As we have seen in Chapter 3 (p. 81), Geoffroy favoured mechanical explanations like Lemery's. He must have been aware of Newtonian explanations of chemical phenomena, but does not refer to them in his own work and presumably does not accept them.

Geoffroy may indeed have been influenced in inventing his table by Newton's brief mention of an order of reactivity in the Queries at the end of the *Opticks*, which we know he had read by 1718; but if so his table is only faintly foreshadowed by Newton, and not anticipated. It is unlike anything known before it in chemistry, and makes the simple description given by Stahl five years later of the 'order in which metals dissolve' seem crude. However, it had many successors during the eighteenth century. It is entitled *'Table des différents Rapports'*, clearly to avoid the connotations of both *'affinité'* and *'attraction'*.

Geoffroy's table consists of sixteen columns, in which the various substances are indicated by the customary symbols, alchemical in origin. At the head of each column is the symbol for the substance or group of substances that it concerns. Below that the symbols for substances with which it reacts are arranged in order of their affinity for it, so that the nearest has the greater affinity for it and cannot be displaced by any of the substances lower down, but can remove any of them from combination with it. This is the normal form for affinity tables for the rest of their history, with few exceptions. It may possibly have been suggested to Geoffroy by the methods of tabulating experimental data which were usual among physicists, or perhaps even by astronomical tables. It was seen as an attempt to give chemistry some of the precision of those more exact sciences.

Table 4.1 shows all the tables of affinity published in the eighteenth century of which I have found mention, with their dates of issue or of publication, together with three which were apparently distributed or shown to students during lectures. It is remarkable that only one table,

Table 4.1. Tables of affinity and similar tables published in the eighteen century

	Date of issue	Author	Description
1.	1718	Geoffroy	16 columns. Uses alchemical symbols.[2]
2.	1730	Grosse	19 columns. Symbols. Derived from Geoffroy.[3]
3.	1749	Clausier	Not in columns, but 78 lists, with about 37 different substances at the heads of them, of substances in the order of affinity. Written in words.[4]
4.	1751	Gellert	28 columns, representing order of increasing solubility, not decreasing affinity. Symbols.[5]
5.	1753	Lewis	12 lines. Words.[6]
6.	1756	Rüdiger	15 columns of substances that combine, and 10 supplementary columns of those that do not. Symbols.[7]
7	1758	Limbourg	33 columns. Symbols.[8]
8.	1762	Marherr	120 columns, with 20 different substances at the head of them, each showing the order of affinity of only two substances with that at the head. Symbols.[9]
9.	1763	Rouelle	19 columns, very little different from Geoffroy's. Symbols.[10]
10.	1763	Spielmann	28 columns, very similar to Gellert's. Symbols.[11]
11.	1769	Warltire	29 columns for 26 different substances or classes. Words.[12]
12.	1769	Demachy	20 columns and 10 appendices. Symbols.[13]
13.	1772	Tessari	17 columns, derived from Geoffroy's. Symbols.[14]
14.	1773	Sage	13 columns, with exceptions at the bottom. Symbols.[15]
15.	1773	De Fourcy	36 columns. No copy survives.[16]
16.	1775	Erxleben	36 separate lists, in words. Derived from Gellert.[17]
17.	1775	Bergman	50 columns, with separate sections for the wet and dry ways. Symbols.[18]
17a.	1783	Bergman	Extension of 1775 table to 59 columns. Symbols.[19]
Variations of Bergman's table			
17b.	1782	Elliot	Shortened version of Bergman's 1775 table, with slight alterations. 50 columns. Symbols, except that substances at heads of columns are in English words.[20]
17c.	1785	Beddoes	Transcription into English of Bergman's 1783 table, with slight alterations.[21]
17d.	1788	Berkenhout	Shortened version of Bergman's 1783 table. 36 columns for wet and 25 for dry way. Symbols, except for headings.[22]

Table 4.1. (*Continued*)

	Date of issue	Author	Description
17e.	1789	Hopson	Derived from Berkenhout. 36 columns for wet and 25 for dry way. Words.[23]
17f.	1790	Richardson	20 columns for wet way and 15 for dry way. Words.[24]
17g.	1790	Nicholson	40 columns, with separate sections for dry way where appropriate. Words.[25]
17h.	1799	Pearson	62 columns, with separate sections for 'in water' and 'in fire'. Words.[26]
18.	1777	Weigel	Not in columns. Distinguishes wet and dry ways and reactions involving two, three, or more substances. Shows products of reactions. Words.[27]
19.	1781	Wiegleb	Not in columns. Distinguishes wet and dry ways and reactions involving two, three, or more substances. Derived from Weigel. Words.[28]
20.	1782	Elliot	16 columns. Words.[29]
21.	1789	Lavoisier	The latter part of Vol. I of the *Traité elémentaire de chimie* is based on 41 tables of combinations, of which 25 are in order of affinity and virtually form an affinity table. Words.[30]
22.	1790	Gergens and Hochheimer	49 columns, distinguishing wet and dry ways where appropriate. Words.[31]
23.	1790	Gren	53 columns in words. Completely revised version in 2nd edn, taking account of Gren's version of oxygen theory; is not in columns; has 11 sections for substances not prepared separately and 76 for those prepared separately, in words, distinguishing wet and dry ways. Abridged version of 1797, in words, in five classes with a total of 60 columns.[32]
24.	1800	Trommsdorff	58 columns, with diagrams of double affinities in addition. Words.[33]
Parts of tables published separately			
25.	1755	Black	Two columns, in words, to be added to Geoffroy's table, following Black's discovery of fixed air (carbon dioxide) and showing the affinities of alkaline substances free of fixed air for acids and for fixed air respectively.[34]
26.	1758	Lesage	One column of affinities for light (or elementary fire) and one column for affinities of glass for different colours of light. Symbols.[35]
27.	1763	Lewis	15 lists, each of three substances, in words, showing the affinities of platinum.[36]

Table 4.1. (*Continued*)

	Date of issue	Author	Description
28.	1764	Meyer	One column for affinities of *acidum pingue*. Symbols.[37]
29.	1783	Lavoisier	Necessary corrections to existing tables to include oxygen and the implications of Lavoisier's oxygen theory. Words.[38]
Unpublished tables found among manuscript lecture notes			
30.	*c*.1757/8	Cullen	24 columns with 18 different substances at their heads. Words.[39]
31.	*c*.1760	Cullen	31 columns. Symbols.[40]
32.	*c*.1767/8	Black	20 horizontal rows for 17 different substances or classes, including two rows for attractions of fixed alkali 'in consequence of heat'. Symbols.[41]

and that no more than a modification of Geoffroy's, was published between 1718 and 1749. Possibly there were others drawn up in this period that have not come to light, but at any rate it is fair to say that if there were any they received little attention. Gellert's table of solubilities, as we shall see, was not really a table of affinities, though it resembles one. Macquer in his influential *Elémens de chymie-théorique* of 1749 simply introduces Geoffroy's table and implies that he was not aware of any others.[42] Then there are three tables in the 1750s, three in 1762–3 (two of them very similar to earlier ones), and a spate in 1769–77. After that it is Berman's table that is dominant.

Why should there have been this long gap between the publication of Geoffroy's table and the renewal of general interest in the subject? The resistance to Newtonian ideas and the loyalty to Cartesianism which held back speculation about chemical attraction in France and in countries under French influence have already been mentioned. Even though Geoffroy himself seems to have avoided the implication that his work was on Newtonian lines, it evidently struck his contemporaries as leading to Newtonian ideas; and it was probably for this reason that little work was published on affinity on the Continent for almost a generation. The long section dealing with affinities in the *Nouveau cours de chymie* is an exception. There may, of course, have been teaching on the subject which does not show up in publications.

Geoffroy's table was presumably less well known in Britain and Holland: and perhaps also the notion of affinity which was thought to be expressed in Geoffroy's table (though he does not use the word himself) was still felt in Britain to be in some way contrary to the notion that chemical combination was due to attraction between particles. At any rate, there seems to have been no British affinity table before Lewis's in 1753. After the work of Stahl and Boerhaave the main lines of progress in chemistry were for a while in practical discoveries and in the understanding of particular substances and reactions. Such was the laborious work of the German metallurgists such as Pott and Eller. By 1749 enough new facts had been accumulated for attention again to be turned to the possibility of finding some general pattern in them. By this time also the Newtonian type of theory had been known, if not accepted, long enough in France for thinking about it to have reached the stage of reconsidering Geoffroy's table.

The efficient cause of the revival of interest in affinity in the 1750s was probably the publication of Macquer's *Elémens de chymie-théorique* in 1749. Macquer was largely responsible for the adoption of the phlogiston theory by French chemists, though his version of it differed essentially from that of Stahl. In his book, however, he lays even more stress on the concept of affinity. The *Elémens de chymie-théorique* was the first of a new kind of textbook which set out to teach chemistry as a science and not as a collection of recipes.

After describing the various classes of substances and their properties, in the last chapter but one of the theoretical volume of the *Elémens* Macquer gives a detailed explanation of Geoffroy's Table of *Rapports*, which he reprints unaltered at the end of the book. Unlike Geoffroy, Macquer has no inhibition about using the word *affinité* with reference to the table. Indeed, he says that since almost all chemical phenomena depend on affinities, the table, which collects together all the facts about affinity, acts as a summary of his whole book. Geoffroy's table, he admits, is open to alterations in many places, as since its author's death new experiments have indicated new affinities and exceptions to old ones; but Macquer's reasons for not making any alterations are that many of these new facts are disputed, and that the existing table gives the fundamental affinities appropriate for elementary work. Most of the rest of this chapter is concerned with the possible exceptions and additions to Geoffroy's table, and it ends with a recommendation to refer constantly to the table while one is reading the book, so as to fix it better in the memory.[43] Macquer's views on affinity will be discussed further in the next chapter.

The explanation for the spate of affinity tables in the period from 1769 onwards is not obvious. Perhaps it is merely a matter of following a fashion which had begun for no particular reason, or conceivably it may have been started by the publication of Macquer's *Dictionnaire de chymie* in 1766. Nevertheless, Geoffroy's table stands out as being well ahead of its time. His idea was still fresh and capable of development sixty years after its appearance.

In Geoffroy's paper introducing the table he simply states the observed fact that, when a number of substances that are mixed together have some disposition to unite, any given one of them always unites with a particular one of the others in preference to the rest. He thus arrives at the following very general proposition:

Every time that two substances which have some disposition to join with each other happen to be united together, if there supervenes a third which has more relationship [*rapport*] with one of the others, it unites with it and makes it release its hold on the other.[44]

This implies the crucial assumption that the *rapports* are constant in all circumstances.

Like many of the later compilers of tables, Geoffroy observes that his table is not complete, but might be added to if further *rapports* were determined experimentally. He adds that in many of the experiments on which his table is founded the separation of the metals is not perfect, because of the unavoidable limitations of experimental conditions; but the errors are not so large as to prevent the rules from being regarded as constant. In the table gold is on its own at the bottom of the second column, which is headed by acid of marine salt (hydrochloric acid), and there are two blank spaces above gold between it and mercury. This represents the observation that gold is soluble only in *aqua regia* (a mixture of nitric and hydrochloric acids), and not in hydrochloric acid alone.

In the second space in the fourth column is the symbol later used for phlogiston, made up of the symbol for sulphur with the three circles representing oil at the corners of the triangle: . In the key at the bottom of the table, this symbol is interpreted as 'oily principle or sulphur principle', and the word 'phlogiston' is not used. As was mentioned above, Geoffroy in his lectures stated that Fire, Water, and Earth were the three elements or principles, that Salt arose from the combination of those three, and that Sulphur or Oil arose from the combination of the three principles with Salt. Phlogiston in Stahl's theory was, of course, derived from Becher's 'fatty earth' (*terra pinguis*), which corresponded with sulphur in the three-principle system. Thus Geoffroy's oily or sulphur principle is similar to Stahl's phlogiston, but not the same. It is shown in the table as having stronger affinity with vitriolic (sulphuric) acid than have alkalis, absorbent earth, iron, copper, or silver, which presumably means that the acid is, in modern terms, a strong oxidizing agent.

The relationship between the 'acid spirits' at the head of the first column and the three individual mineral acids at the top of the next three columns is not explicitly stated. However, the context of the columns makes it clear that the first refers to the affinities of all three mineral acids with alkalis, metallic substances in general coming at the bottom of the column, whereas in the next three columns the differing affinities of the three acids for the various metals (and in the case of vitriolic acid for the sulphur principle) are dealt with in detail. Geoffroy's symbol for acids in general, which is not one of the traditional symbols, is like a capital omega (↶·↷) with arrow heads at the ends of the lines at the base of it. Probably the arrow heads represent the spikes on the particles of the acids, to which acid-properties were due according to the belief of seventeenth-century mechanical philosophers, supported by Geoffroy in his lectures.

Similarly the 'metallic substances' symbolized as a general class at the head of the eighth column are apparently the same as those dealt with separately in the tenth to fifteenth columns. In the eighth column their affinities for the acids (including vinegar) are dealt with, and in the succeeding columns their affinities with each other. Their combinations with each other would, of course, take place only 'in the dry way', that is at high temperatures, so that even in this early table there is an implicit distinction between affinities in the wet way and in the dry way.

It is noticeable that except for the inclusion of vinegar in the column for combinations with fixed alkali and metallic substances, and for the last column, which shows that salt is removed from combination with water by spirit of wine, the table is entirely concerned with mineral chemistry. The substances chiefly included are the three mineral acids, the common alkalis, metals, and a few earths. In 1718 this covered the normal range of chemistry. The composition of organic substances was hardly understood at all. Indeed, although a number of organic acids were newly identified by Scheele and others later in the century, even by 1800 no more than a start had been made (through Lavoisier's work) on discovering their constitution.

In columns 14 and 15 silver, copper, and lead occupy the same square. That evidently means that each can combine with the substance at the head of the column (iron and antimony respectively), but that none will displace any of the others from combination with it. This device is not uncommon in later tables.

The trouble with affinity tables was that they represented too simple a view of chemical combination. Too many factors are at work in determining how a reaction is to proceed for such an uncomplicated scheme to show them. Exceptions were soon pointed out—that is, reactions in which the normal order of affinities seemed to be reversed. Reversible reactions were also pointed out. Then it was realized that most reactions involved more than two substances, so that the result would depend on the algebraic sum of competing affinities if it depended on such constant factors at all. Further complications were the effects of heat and of differences in the concentration of the reagents, as well as differences in solubility or volatility. Several attempts were made to allow for these complexities, though inevitably they were in the end unsuccessful.

Even as early as 1720 Geoffroy the elder had to defend his table against certain objections, of which two (according to Demachy) were made by Geoffroy's younger brother and the third by Neumann, chemist to the King of Prussia.[45] The first objection was that although the table showed a greater *rapport* between acids and alkalis than between acids and earths, in fact quicklime (calcium oxide) detaches urinous acid (i.e. ammonia) from sal ammoniac (ammonium chloride). Geoffroy the elder replied that although limestone (calcium carbonate) was an absorbent earth, quicklime was a different substance and was virtually an alkali.

The second objection was that whereas the table showed closer *rapport* between acids and alkalis than between acids and metallic substances (in the first column) in fact metallic substances sometimes detached acids from alkalis. For instance, in the preparation of *Fleurs Martiales* (ferric chloride) from sal ammoniac (ammonium chloride), iron filings are mixed with a little of the salt and heated. At the start of the reaction a little ammonia is carried over which can have been detached only by the iron from the acid to which it was united. Therefore in this case the iron, though a metal, has more *rapport* with the acid than has the alkali. Geoffroy the elder replied that iron is not quite the same substance as iron filings, which are to some extent decomposed by having been digested for twenty-four hours for this experiment.

The third objection, made by Neumann, was that if three parts of minium (lead oxide) and one part of sal ammoniac were mixed and distilled together, a good quantity of very strong spirit of urine (ammonia) was produced. This implied that the minium detached the spirit of urine from the marine acid (hydrochloric acid) contained in the sal ammoniac, and so had the stronger affinity for the marine acid. The elder Geoffroy in his reply added that he had found that the effect was the same with either of the calces of lead (i.e. litharge as well as minium) and with the calces (oxides) of tin and antimony. Minium was, after all, only a calx of lead 'reverberated until it has acquired the red colour'. Although he admitted that in this reaction the metallic calces detached the spirit of urine from the marine acid, he claimed that they did that only by the intermedium of a fixed alkaline salt which they contained, and which was discovered or formed during the calcination. Calx of lead was a compound very different from lead.

Geoffroy here seems to imply that the calx contains something which the metal does not, which is quite the opposite of what adherents of the phlogiston theory believed, when it came into favour. In his replies to the objections Geoffroy also implies that each compound has different affinities—for instance, that limestone and quicklime have different affinities, even though they presumably have at least a component in common, that iron and iron filings have different affinities, even though their composition must be very similar, and that calx of lead and lead have different affinities. Early in the eighteenth century it was still normal, at least on the Continent, to assume that compounds showed properties which were merely a blend of those of their components; but consciousness that compounds should be regarded as distinct from elements, and so the tendency to refer affinities to elements, or at least radicals, developed later in the century. By the third quarter of the century, it was becoming more natural to refer all affinities to the simple components of compounds, and to suggest that if the fact that the components were combined made any difference to their affinities, it merely modified the essential affinities and did not require the chemist to consider the affinities of the compound as a separate case. Geoffroy, however, does not think of affinity as being a property of the elementary components, but as a property of the whole compound.

Long after the elder Geoffroy's death, in the volume of the *Mémoires de l'Académie Royale des Sciences* dated 1744 but not published until 1748, his younger brother respectfully pointed out another exception to the table, in a memoir entitled 'Observations on the earth of alum [aluminium oxide]: method of converting it into vitriol [here aluminium sulphate], which makes an exception to the Table of Relationships in chemistry'.[46] He had found that although the table showed absorbent earths as having a stronger *rapport* for vitriolic (sulphuric) acid than had metallic substances, a solution of alum (a double sulphate of aluminium and another substance) dissolved a certain proportion of green vitriol (ferrous sulphate) without going cloudy, and then that iron would precipitate the earth of alum. It followed that the earth of alum had a weaker affinity with vitriolic acid than had iron, and a correction should be made to the table. Geoffroy adds the wish that all chemists would look for similar exceptions, 'so that one could render still more perfect this Table of relationships which is already so useful to Chemistry'.

An impressive feature of this memoir is that Geoffroy reports weighings which show that the loss of weight of the iron by solution was equal to the weight of the earth of alum which it precipitated. Obviously he has too simple a notion of the connection between the two weights, and the equality is just a coincidence. Nevertheless, it was still not common at this date to use quantitative evidence of this kind in chemistry at all. In this memoir phlogiston is mentioned, and the words *rapport* and *affinité* are used as equivalents. Evidently by the 1740s the elder Geoffroy's careful avoidance of the word *affinité* was no longer necessary, as it had lost much of its connotation.

Grosse and Clausier

Grosse's table of 1730 is printed along with several others by Demachy at the end of his *Recueil de dissertations*, and he says that Grosse was a recluse whose table was made public only by his pupils.[47] It is an extension and revision of Geoffroy's table, showing the following main differences:

1. A column for *Aqua Regia* is added after that for vitriolic acid.

2. The columns for sulphur and mercury are placed before that for Metallic Substances in general, instead of after as in Geoffroy's table. Presumably the reason for the change is that Grosse still thinks of sulphur and mercury as being principles, in the Paracelsan tradition, and therefore not on the same footing as the metals lead, silver, copper, iron, and antimony.

3. Instead of showing a single constant order of affinity with metallic substances for all the four acids concerned, as did Geoffroy, Grosse merely shows marine acid (hydrochloric acid) as coming before acid spirits in general in affinity for metallic substances, and then shows in the same column three separate examples. Marine acid has a greater affinity for lead than vinegar, for mercury than nitric acid, and for silver than vitriolic (sulphuric) acid.

4. Grosse shows copper above silver instead of below in the column for lead.

5. Grosse shows the column for silver before that for copper instead or after.

6. In the column for water immediately following that for regulus of antimony (metallic antimony), Grosse puts salt above spirit of wine (alcohol) instead of below.

7. Grosse adds a short column for tartar (potassium hydrogen tartrate).

8. After the column for tartar, Grosse adds a second column for water, in which spirit of wine is shown above oil. Several later tables similarly have two columns for the same substance, showing the order of affinities in a different series of reactions.

9. Grosse has inserted some additional substances in the first ten columns. One of these (in columns 6 to 8) is symbolized by the usual symbol for sulphur with a 'v' on its side in the lower left-hand corner. This horizontal 'v' usually means 'acid', and the compound symbol presumably means 'acid of sulphur' (sulphuric acid).

The third and seventh differences are explained by two memoirs by Grosse, the former in collaboration with his colleague Duhamel.[48] With Duhamel he investigated the nature of tartar, and gave in words a much fuller list of its affinities than he gave in his table. In the second memoir, on lead, he said that he had confirmed Geoffroy's Table by finding that mercury should go below lead in the column for nitric acid, but found an exception to Geoffroy's eighth column which showed marine acid (hydrochloric acid) as having more *rapport* with Metallic substances than had other acids. In fact, vitriolic acid (sulphuric acid) had 'a strong relationship' (*beaucoup de rapport*) with lead.

Many, though not all, of the differences between successive tables were of this kind—new substances added, slight changes in the order of substances in particular columns, representing a different interpretation of the same experiment or a preference for the results of a new experiment rather than an old one, or differentiation between substances which had previously been classed together. It will not always be necessary to comment in detail on such differences.

Clausier, who translated Quincy's *Pharmacopoeia* into French in 1749, inserted his own table of affinities and replaced Quincy's theoretical section by a long description of his own theories, which have been briefly described in the last chapter.[49] The 'Table of sequences of affinities' (*Table des suites d'affinitez*) is written in words, like a number of later tables, not in symbols, 'for the convenience of those to whom these characters are not familiar', and is printed continuously and not in tabular form. Some of the sections are very general. The first, for example, entitled 'General Affinities' (*Affinitez générales*), merely states that 'Substances attach themselves to those which are like them.' 'Like' (*semblables*) is not here used in a superficial sense, since metals are shown as combining with acids, and so forth. Clausier is referring rather to a supposed correspondence between the shapes of their pores, as he explains in his text. The second section, entitled 'Particular' (sc. affinities) consists of the statement that 'Fire is attracted and repelled by the vibrations of the walls of substances, all of which it dissolves, except earth.' There are also sections for 'salts in general', 'acids in general', and 'white metallic substances', as well as the sections dealing with particular substances of those classes.

The whole table falls into two parts, Sections I to XLVII headed 'Principally solvents', and Sections XLVIII to LXXVIII headed 'Affinities of metals'. Although the fact is not explicitly stated, the first part must refer to reactions at ordinary temperatures, 'in the wet way', and the second to reactions at high temperatures, 'in the dry way'. Like Geoffroy's separation of the affinities of metallic substances for acids from the affinities of particular metals for each other, this division faintly foreshadows Baumé's suggestion, later carried out by Bergman, that there should be two tables, one for the wet way and one for the dry way.

There are seventy-eight sections altogether. However, some substances head several different lists, representing different series of affinities which could not be compared because they would not be observed in the same sequence of replacement reactions, like Grosse's two separate columns for water. There are thirty-seven different substances at the heads of sections, not counting 'acids in general', 'salts in general', or 'white metallic substances' (as classes of which particular members are later dealt with in separate sections), but counting *acide du nitre* and *eau fort* (both presumably niric acid) as distinct, and including some neutral salts which do not usually occur in affinity tables because they are compounds of acids with alkalis or metals, and their reactions are usually supposed to depend on the affinities of their components. However, sections XXVI to XXX do not really show affinities, but just properties by which salts may be identified, such as 'Vitriol of copper [copper sulphate] with gall-nut gives a greenish blue colour.' In the part of the table dealing with neutral salts, only Sections XXXI to XXXIII show affinities, and those are headed by salts which are occasionally shown in their own right in other tables, viz, marine salt, tartar (potassium hydrogen tartrate) and sal ammoniac (ammonium chloride).

Often more than one substance is shown as having equal affinities, usually with commas between them, as in Section XIII, no. 2, where 'iron, copper, lead, zinc, regulus of antimony [metallic antimony], chalk, coral, marble, and pearls' are shown as having affinity with distilled vinegar (acetic acid) without any order being implied. In one case, Section L, no. 2, the word 'or' is used ('sulphur, copper, or lead'); but no doubt it is implied in all similar cases. Sometimes two substances are joined by the word 'and', which means that they act together in the same replacement reaction, as in Section XLIX, where 'air and fire' are shown as having closer affinity with gold than has regulus of antimony or earth. Sometimes at the end of a numbered sequence of affinities one or more substances are shown without a number, which means that they combine with the substance at the head of the sequence but that it is not determined whether they have greater or less affinity with it than have the numbered substances. For instance Section V shows absorbent earth, sulphur and earth together, and sulphur alone (in that order) as combining with salts in general, and adds 'The base of the food (*aliment*) of fire' separately. In one or two sections there are only two substances, meaning no doubt that the two combine together and not implying that any other is replaced, as in Section LXIX. It shows lead as combining with copper, in contrast to LXX, which shows mercury as displacing calamine (zinc carbonate) from combination with copper.

In some of the sections (for instance no. LXVII) the separation of two substances by fire is shown by putting them as first and third in a sequence of affinities and fire as second. This is unusual, apparently because chemists did not usually count fire as a chemical substance even though at this period they mostly thought of fire or heat as a material fluid. A little later, indeed, there was a theory that phlogiston was fire which had entered into chemical combination; but Clausier evidently does not hold that view.

Many of the substances which he includes, particularly organic substances of which the composition was not then understood, such as 'saliva', 'mucilages', or even 'heterogeneous substances', would not now be thought of as distinct chemical species. However, the same is true of a number of other tables, even those of Weigel, Erxleben, or Gren a generation later—in fact, the inclusion of a substance is evidently not intended to imply necessarily that it is philosophically simple or elementary, but merely that it is referred to in ordinary chemical practice as a reagent. After all, one of the chief tasks of chemists during the eighteenth century was to establish a working list of substances that could usefully be treated as simple or elementary for the purpose of understanding chemical composition and reactions, independently of traditional philosophical beliefs which were not useful for that purpose. In the process of developing such a working list it is not surprising if some chemists tried to include substances which seem inappropriate to us with hindsight, and which were later found not to be helpful items on the list and dropped from it. It must also be remembered that there was then little knowledge of the composition of organic substances, so that they could only be included in tables by referring to vague general categories. Some of Clausier's items are rather extreme examples.

His table was criticized as vague and unsatisfactory by both Demachy and Guyton de Morveau.[50]

Gellert and Lewis

The textbook of metallurgic chemistry of C. E. Gellert (1713–95) is a typical example of a textbook of the period before the appearance of Macquer's *Elémens*.[51] Most of it is devoted to descriptions of apparatus, practical processes, and the various substances and ores. The space given to theory is still comparatively small. The interest in metallurgy and its practical applications is characteristic of German work at this time, and, as Demachy remarks, Gellert has used Pott's work on the classification of 'earths' according to their fusibility.[52] Five earths are included, as against Geoffroy's single 'absorbent earth'. Gellert is often referred to by Macquer, and his table is mentioned by Baumé along with Geoffroy's.[53] He fully accepts the notion of phlogiston, but follows Boerhaave's views on heat and its operation. He attributes the solution of salts in water to attraction.

Occasionally he speculates about the size of particles, suggesting for instance that although the particles of air are so small as to be indiscernible by microscopes, they must be larger than particles of fire, because they do not seem to penetrate solid bodies which are penetrable by fire.[54] Presumably this particular speculation, which would have seemed rather old-fashioned in Paris, was permissible to an orthodox chemist because it did depend on experimental evidence.

Demachy complains of Gellert's table that he has unnecessarily reversed the usual order by putting substances that have the least *rapport* with the substance at the head of the column nearest to it.[55] That is a misunderstanding. Gellert's table does not claim to be a table of *rapports* or affinities, but only to show which substances are soluble in the substances symbolized at the head of each column. Admittedly, eighteenth-century chemists often saw solution as a kind of chemical combination; but even so solubility was not quite the same as affinity. The symbols in each column are arranged so that any substance will be precipitated from solution by any of those below them. Thus the order in each column is naturally the opposite to that which would be found in a table of affinities. In fact, although the table is usually included by later writers among affinity tables, it shares only a part of their purposes. It is meant primarily to be of practical use in following industrial processes, and not to lead on to any theoretical implications. Although it has no column for water, it includes three metals or semi-metals that are not in Geoffroy (cobalt, arsenic, and bismuth) as well as nitre, liver of sulphur (polysulphides of potassium), glass of antimony (antimony trisulphide partly converted to antimony trioxide by roasting), and glass, and gives zinc a column of its own.

Gellert mentions in a note that the precipitation implied by the table does not always take place for two reasons:

> 1st. Because this dissolving body will either dissolve one of them only in a small degree better than the other; or 2dly, because often those which shall be dissolved or precipitated are liable to dissolve one another themselves; now and then both reasons will occur at the same time.[56]

Thus Gellert has picked out two of the factors which made it difficult to derive from experiment a single order of affinity which could be relied on in all circumstances.

The column at the right-hand side of the table shows glass as combining with the calces (generally oxides) of metals, not with the metals themselves. That no doubt represents the practical reality; for glass would be coloured by the calces of metals (that is, oxides or other naturally occurring compounds) rather than by the metals themselves. In terms of the phlogiston theory it would also be logical to show the calces rather than the metals in a table, for the calces were supposed to be simple substances and the metals themselves compounds of their calces and phlogiston. However, that may not have been in Gellert's mind, for he shows the metals and not their calces in all his other columns. It was left for Bergman to take the theoretically correct step and show the calces in place of the metals in all columns.

At the bottom of seventeen of the twenty-eight columns of Gellert's table are the symbols for some substances which are completely insoluble in the substance represented at the head of the column, except that where silver is given as completely insoluble in fixed alkali Gellert adds the note 'yet this but partly in the dry way'. Also in his remarks on the table Gellert says that

> It is a matter of great difficulty, chiefly in the dry way, to ascertain the order in which bodies may be dissolved; Whence it must needs remain liable, here and there, to some objections. Nevertheless it will be found much compleater than any that has appeared before of this kind.[57]

Here is a hint of the necessity for defining the temperature of a reaction when quoting affinities, and to a very slight extent Gellert anticipates Bergman's separation (already foreshadowed by Geoffroy and Clausier and later explicitly suggested by Baumé) of the table of affinities in the dry way from the table of affinities in the wet way.

In remarking that his table 'will be found much compleater than any that has appeared before of this kind', Gellert is of course conceding that his work is related to that of Geoffroy, as indeed is obvious from the very nature and appearance of his table. However, he says nothing about his reasons for altering the basis of the table and making it into a table of solubilities. We can only infer that he is displaying a refusal to take any philosophical stance on the nature of affinity, or the validity of the concept of affinity, as some later compilers of tables were to do, and simply avoiding that side of things altogether and recording the observed data in a form which will be of practical use to metallurgists.

As Dr Sivin has pointed out in his penetrating study,[58] William Lewis was also concerned with the application of chemistry to industry as well as with pure chemistry, seldom discusses philosophical theories of chemistry in his writings, and makes it clear that his table is presented for its practical value rather than for any theoretical implications. Elsewhere, in his *Commercium philosophico-technicum*, Lewis explains his views on the operation of chemical attraction, which are normal for a British chemist of his time.[59] His distinction between 'mechanical attraction' and 'chemical attraction, or affinity' has been quoted in Chapter III (p. 102).

The title of his table is 'A Table of the relations or affinities between different Substances', 'relations' being presumably a translation of Geoffroy's *rapports*; but Dr Sivin has pointed out a number of differences from Geoffroy. These may be summarized as follows:

1. Lewis's table is in words, and has the substances which combine in horizontal rows instead of columns.

2. Lewis has omitted the sequences of affinities for mercury, lead, copper, silver, and iron, which refer of course to affinities 'in the dry way', that is to reactions at high temperatures between molten metals, rather than to reactions in aqueous solution at room temperature, which

are more likely to be relevant to pharmacy. Lewis has also inserted a new line for 'inflammable spirits' at the beginning of the table. The new sequence corresponds with Clausier's no. XLV, for Lewis's phrase 'inflammable spirits' means the same as the French 'esprit de vin', that is spirit of wine or (in modern terms) ethyl alcohol. Since Lewis's book is intended as 'a Correction, and Improvement of Quincy', it seems quite likely that Lewis had seen Clausier's translation of Quincy with its table.

3. Lewis has altered the order of the sequences, putting the sequence for water before that for acid spirits and immediately after his new one for 'Inflammable spirits', and that for 'alcaline earths', which corresponds with Geoffroy's 'absorbent earth', after those for fixed and volatile alkaline salts (the hydroxides of potassium or sodium and of ammonia respectively).

4. Lewis has added a second sequence for water to that which he has taken from Geoffroy. The new sequence shows 'fixt alcaline salts' as having a closer affinity with water than have inflammable spirits, rather like the column included by Grosse in his table in which he shows salt above spirit of wine instead of below it as in the previous sequence and in Geoffroy's version. Dr Sivin suggests that Lewis has inserted a second sequence to support a point that he made in discussing the crystallization of salts:

> Some even of the neutral kind, particularly those of which certain metallic bodies are the basis, are so strongly retained by the aqueous fluid, as not to exhibit any appearance of crystallization, unless some other substance be added with which water has a greater affinity.

However, this point refers to the former of the two sequences for water, which has been taken from Geoffroy, who made exactly the same point. Since 1718 the distinction between neutral salts and alkaline salts (which we should not nowadays call salts at all, but just alkalis) had been generally accepted, and Lewis has added the second sequence and the word 'neutral' to the word 'salts' in the former sequence simply to indicate that the two classes of salts behave differently. Presumably Grosse even in 1730 had the same distinction in mind.

5. Lewis has used slightly more general terms than Geoffroy, for example, 'alcaline earths' for 'absorbent earth', 'inflammable spirits' for 'spirit of wine', and 'Vegetable acids' (i.e. organic acids) for what Geoffroy called merely 'spirit of vinegar'. The reason, again, is evidently that since 1718 more than one species had been distinguished within these classes, and Lewis wishes to include the whole class.

6. Although Lewis does not alter Geoffroy's order within any sequence, he adds some substances (for example 'Oils' at the end of the sequence for metallic substances, and 'camphor', which is also in

Grosse), at the end of the sequence for nitric acid. He also omits some: for example gold from the sequence for sulphur, in accordance with his view, expressed in his edition of Neumann, that gold does not react with sulphur.

Dr Sivin suggests that the sequences for metals which Lewis omits are a weak point in Geoffroy's table, since if it were not for the effects of mass-action in particular cases they would all be the same and could be shown in a single generalized column; but since particular conditions do in practice govern particular cases the sequences for iron and for regulus of antimony (metallic antimony) in Geoffroy's table appear to be inconsistent. The criticism of Geoffroy is perhaps a little unhistorical. No doubt if Geoffroy's table were to be published today it would be helpful to point out such weaknesses: but we must look at it as the very first table of its kind, published in the state of understanding of its time. Seen in context, it is bound to show sequences of reactions as they were observed then, and could hardly reflect reaction mechanisms as illuminated by late-nineteenth-century thinking. This kind of problem was, of course, continually arising in affinity tables, and most compilers tried the same kind of alteration as Lewis, in unsuccessful attempts to produce an order which was universally applicable.

In his notes in his edition of Neumann's works of 1759 Lewis by implication extended the sequence of affinities for various substances beyond that shown in his table.[60]

Platinum was made known to eighteenth-century chemists by William Brownrigg (1711–1800) in papers read to the Royal Society in 1749 and 1750; but Lewis was the first to give it the status of a new metal, and in his *Commercium philosophico-technicum* he described it thoroughly and produced fifteen sequences, each of three substances, and including platinum in each case, to show its affinities. Each sequence shows either that one substance displaces another from combination with platinum, or that platinum displaces another substance from combination with a third or is displaced by it; and the experimental evidence is quoted for each sequence.[61]

Cullen and Black

There are a number of manuscript copies extant of notes taken from the lectures on chemistry given by William Cullen (1710–90) at Edinburgh, which were a new departure in that they were delivered in English instead of in Latin. The lectures of his predecessor Plummer were in Latin, as was normal in European universities. Cullen lived, of course, in the period of great brilliance of Edinburgh's intellectual life now known as the Scottish Enlightenment. A number of courses of elementary lectures at this time included an exposition of Geoffroy's table, which was as Macquer had pointed out a handy way of clarifying the main outlines of chemistry for the beginner. However, at least two of the manuscripts of Cullen's lectures contain affinity tables of his own, together with explanations of them. The manuscripts seem to be fair copies rather than notes actually taken down in the lecture room, and we cannot be sure that the version of the tables is correct in every detail, or indeed whether they were dictated or copied down from a master written version.

The earlier of the two tables, that included in the manuscript belonging to Dr W. A. Smeaton, which is probably to be dated about 1757–8, is in words tabulated in columns. However, it is possible that the student who wrote the notes may have transliterated the table from symbols. Twenty-five columns are given, but no. 7 at the left of page 635 and no. 8 at the beginning of the next page are the same. Analogy with other tables suggests that the second may have been copied in mistake for a column headed by volatile alkali (ammonia). As Cullen's detailed explanation makes clear, the table is based on Geoffroy's, but much has been added in the light of experiment. A column for fixed air (carbon dioxide) has been included, with acknowledgements to Cullen's pupil Black. At the start of the explanation of the table (p. 638) Cullen says:

> In Gellot's [*sic*] Metallurgy there is a Table of Attractions, but in that the Intermediate Substances will not separate [the lower ones], hence for distinction's sake, here we use the term of Elective Attractions.

Evidently Cullen shared the common misunderstanding of Gellert's table of solubilities.

Cullen was probably the first to use the term 'elective attractions', though it is just possible that he may have got it from his pupil and successor Black. Cullen normally speaks of 'attraction', as was usual in Britain, but he occasionally uses the word 'affinity' (pp. 654, 657), evidently with no difference in meaning. At any rate, he has felt no need to comment on the difference between his own name for Gellert's table ('Table of Attractions') and the one usual in earlier works ('Table of Affinities'). Elsewhere (on p. 112 of the Wellcome Library's manuscript) he speaks of 'repulsion'. Cullen's use of an ether theory—an unusually speculative line of thought for a chemist of his time—has been mentioned in the previous chapter. In the course of the explanation of his table he uses a number of the diagrams with four symbols at the corners of a square and arrows along the diagonals to illustrate double decompositions, which he attributes to 'double elective attractions' (see pp. 145ff). He also refers to Lewis's table.

The table in the manuscript in the Wellcome Historical Medical Library has thirty-one columns, with twenty-six different substances at their heads. It is an enlargement of the earlier one, and may be tentatively dated about 1760. It uses symbols, some of which are peculiar to Edinburgh.

The table in the notes of Black's lectures taken by Thomas Cochrane in 1767/8, which have been published by Professor McKie, is in symbols and resembles Lewis's table in that the sequences run horizontally instead of vertically.[62] (This format may, of course, be due to Cochrane's copying rather than to Black.) The table has twenty rows with seventeen different substances at the heads of them, and is essentially a simplified version of Cullen's table. However, it has two interesting features. The normal table is followed by twelve of Black's diagrams illustrating double decompositions, which are not unlike Cullen's but differ from them in having no arrows, which might imply a mechanical interpretation. Bergman greatly extended this series of double decompositions, and the diagrams will be discussed more fully when Bergman's work is described.

Black's table is also remarkable in a second way. Not only are the attractions of the metals with respect to one another, which would presumably be observed in the dry way, shown in a separate division from attractions relating to the wet way, and 'elective attractions in consequence of heat' separated from the 'relation bodies have to water' in the section for single elective attractions; but also in the section for the diagrams of double elective attractions those 'which happen in mixtures of watery solutions' are separated from those 'which happen in mixtures by fusion'. Thus Black anticipates not only Bergman's diagrams of double decompositions, but also to some extent his separation of reactions in the dry way from those in the wet way, which had been recommended by Baumé in 1763. There seems to be a strong probability that Bergman had had some account of Black's lectures, though the evidence is only circumstantial.

Rüdiger, Limbourg, Marherr, and Rouelle

Anton Rüdiger (1720–83) put his finger on the flaw in earlier affinity tables when he wrote that they are lacking because they deal only with affinity, and do not deal with the different circumstances which produce the combinations and precipitations attributed to affinity.[63] However, as Demachy says, he did not succeed in putting that right in his own table.[64] The chief novelty in it is the special little section of ten columns at the bottom of the table showing substances which do not combine, recalling Gellert's section for insoluble substances. Unlike earlier tables, Rüdiger's has the metals at the beginning, followed by the acids, with the alkalis after the acids. The tendency to emphasize the importance of the metals is a German characteristic at this period. Rüdiger has added arsenic to Geoffroy's list of metals, as had Gellert; but not the other substances which Gellert had added. However, he has retained the column for water; included a column for spirit of wine, like Clausier and Lewis; added a column for 'acid of alum' that contains only iron and absorbent earth in that order; and combined the columns for fixed and volatile alkalis into one. ('Acid of alum' is of course the same as sulphuric acid, but was incorrectly taken to be a distinct substance.)

The table of Limbourg, whose eclectic theory of affinity has been described in Chapter 3 (pp. 70–2), consists of thirty-three columns, and so is larger than any of its predecessors except Clausier's, which is not exactly comparable.[65] It resembles Clausier's in including some rather vague substances, such as soap, and has added columns for amber, ether, and orpiment (arsenic trisulphide), as well as retaining Gellert's liver of sulphur (with a modified list of substances which combine with it), together with arsenic, bismuth, and zinc, but not cobalt. Phlogiston is shown at the head of a column for the first time, unless we count Clausier's 'base of the food of fire'. Limbourg does not show all Gellert's earths, but as well as absorbent earth he has aluminous earth, lime, and slaked lime. Presumably he has studied the work of both Gellert and Clausier, whereas Rüdiger may have known only Geoffroy and his countryman Gellert, and added a column for spirit of wine on his own account. Limbourg's inclusion of such substances as aluminous earth, lime, and slaked lime may reflect the growth in interest in the processes of industrial chemistry in mid-seventeenth century France and the Low Countries.

The table composed by P. A. Marherr (1738–71), which is the only one known to me from the Austrian Empire, is part of his doctoral dissertation submitted to the University of Vienna in 1762. In his preface he says that the table of Geoffroy, and other contributions made since, are all faulty to some extent, and that he is setting out to correct the faults.[66] (Twenty-four-year-old doctoral candidates are often like that.) The actual text of the dissertation is entirely concerned with particular cases. Affinity is simply taken for granted, and its nature and cause are not discussed. Most of the space is occupied with the exposure of what he sees as errors in Geoffroy's table—that is, with displacement series which Geoffroy, on the basis of one set of experimental results, has put in one order, but which Marherr, on the basis of a different set of experiments, would put in another order.

He discusses Geoffroy's defence of his table in the paper of 1720, disagreeing with the distinction between limestone and quicklime, and concluding that the question whether that substance or ammonia has a closer affinity with the acid contained in sal ammoniac is 'doubtful' (*anceps*).[67] Marherr was, in fact, the first to use the term 'reciprocal affinity' for the affinity between two substances which could either combine together and displace a third from combination with one of them, or (presumably in other conditions) be separated by the intervention of the same third substance—in other words, for the affinity between substances involved in a reversible reaction. In another passage he disagrees with Geoffroy's placing of phlogiston above ammonia in the column for vitriolic acid.[68]

It is here that he sums up his attitude to chemical method: 'But I require from a chemist not words but deeds; nor do I have any faith in words, but in experiments.'

Marherr's own table is quite different from all the others. Each of the 120 columns contains the symbols for only three substances, and shows that the second substance displaces the third from combination with the first. This arrangement requires much more space than the usual one, as many repetitions of the same symbol at the head of several columns are required to convey the same information as Geoffroy's table conveys in a single column, though admittedly with more opportunities for exceptions. The object is to be able to show affinities which are different in different conditions, especially reciprocal affinities. These are represented by many pairs of columns in which the second and third substances are the same but in reverse order. However, Marherr gives no indication of the conditions to which the columns relate, and does not explain his table in detail in his text. In fact the table is almost a *reductio ad absurdum* of affinity tables. It implies that understanding of the factors at work in directing the course of reactions was quite insufficient to allow them to be represented in a systematic and explicit table.

Marherr refers not only to Geoffroy and his brother, Neumann, Macquer, and Margraf, whom one might have expected, but also to Plummer's papers in *Essays and observations*.[69] Not only was Marherr thorough and industrious, but the volume of current chemical writing in Europe was small enough for one student to master.

As compared with Geoffroy's table, Marherr's table has additional columns for *aqua regia* (a mixture of hydrochloric and nitric acids), phlogiston, vitriolic ether (i.e. ordinary diethyl ether), tin, zinc, liver of sulphur, and gold, but has no column for acid spirits in general, absorbent earth, or volatile alkali. However, mineral acids in general, absorbent earths (plural), volatile alkali, and various other substances which are not in Geoffroy's table are included in the body of Marherr's. On the whole it seems likely that he had used Gellert's table as well, and probably some of the others then available.

The table reproduced by Guyton de Morveau in his *Elémens de chymie, théorique et pratique* in 1777 with the title 'Table des Affinités' is attributed by him to Rouelle.[70] Its first publication seems to have been in the plate headed *Chimie, laboratoire et table des rapports* in the *Encylopédie*.[71] The only differences from Geoffroy's table are the division of the column for metallic substances into two columns for solar and lunar metallic substances respectively, in the former of which the order of vitriolic and marine (sulphuric and hydrochloric) acids is reversed, the addition of two short columns for spirit of wine, the addition of vinegar, acid of tartar (tartaric acid), and sulphur to the columns for absorbent earth and volatile alkali, and the addition of acid of tartar and sulphur to the column for fixed alkali, in which Geoffroy had already included vinegar. These changes are marked with asterisks in both publications.

The table published by J. R. Spielmann (1722–83) of Strasbourg differs from Gellert's in being headed *Table des Rapports* instead of being a table of solubility (although like Gellert's it shows substances in the reverse of the usual order); in omitting the section for insoluble substances and the comments which Gellert added in several places to show that substances were only partly dissolved; and in adding silver in column 9, sulphur in column 6, and two minor notes.[72] In fact, it spoils the point of Gellert's table without substantially improving it.

Tables of 1769–75

The period 1769–75 produced more tables than any other short period, though it cannot be said that any before Bergman's was strikingly original. They are constructed on much the same principles as Geoffroy's over fifty years before. They include a certain amount of detailed information which had been acquired in the mean time; but they do not contrive to take into account the various complexities in the operation of affinities which had been pointed out, for instance, by Macquer and Baumé, though their authors often discuss such complexities. Up to Bergman's time the number of columns in them is not vastly greater than the average of earlier tables, though there is a small increase. Warltire's table is unusual for its time only in being written in English, and in including a column for 'mephitic air' (i.e. carbon dioxide, Black's 'fixed air').[73] It also recalls Black by being headed 'Table of Specific or Elective Attractions', whereas Continental tables still generally at this period have *rapports* or *affinités* or *verwandtschaften* in their titles. Nevertheless, the meaning is the same: the difference is merely a relic of old battles.

Demachy's attack on the whole notion of affinity as the explanation of chemical union has already been described. He also criticizes several existing tables in detail. For example he writes of Geoffroy's table:

> Preoccupied, as were both the Author of the table, and his brother, and their friend [i.e. Neumann] with the assumption that more or less affinity was the sole cause of the phenomena reported in the table of relationships, they did not answer the objections which invalidated this idea, and nothing more satisfying was put in its place.[74]

Demachy must be aiming at interpreters of Geoffroy's table and other tables, for Geoffroy himself avoided mention of anything like affinity or attraction. The tables themselves, apart from any interpretation, may be taken simply as a summary of the facts of chemistry, and are not open to Demachy's attack. Indeed his own table, which is naturally headed *Table des combinaisons* without mention of anything like affinity, is not very much different from most of the others. The detailed exposition of the table and of the set of experiments on which it is based is sound but not outstandingly original.[75] One valuable point is the distinction between natural and artificial combinations, which recognizes the need for specifying the conditions of a reaction when determining affinity. Nevertheless, it undermines Demachy's case, for in referring to one kind of combination as 'natural' he implies that it is this kind which shows the genuine affinities, and that the artificial kind are just a variant.

In any case, by including several substances in a single column of his table Demachy is implicitly admitting that there is something which is a constant factor in a number of different reactions, whatever it is to be called and whatever may be known or unknown about its cause. For a column which includes more than three substances necessarily implies that the order of strength of tendency to combine is the same for a number of different reactions.

The column for the affinities of fire is unorthodox. Although fire was one of the Aristotelian elements, nearly all chemists at this period had a rather more sophisticated view of combustion, and considered it as release of phlogiston. Even heat, though many chemists regarded it as a substance, is not very often included as such in affinity tables or in other lists of simple substances. Demachy in fact puts phlogiston at the top of the column for substances which combine with fire, followed by a number of metals. He may mean by 'fire' very much the same as others meant by 'heat'. His column for fire is also strange in having copper both near the top, immediately under phlogiston, and at the bottom, presumably representing two different reactions.

Apart from this column for fire, the chief novelty in Demachy's table is that it has ten appendices to various columns, as well as its twenty main columns. However, these perform the same function as the columns of exceptions, or the two or more parallel columns for the same substance, which occur in earlier tables.

Tessari's table seems to be the only one published in Italy, but it is no more than a modification of Geoffroy's.[76] The differences are few.

1. Tessari has inserted fixed alkali (potassium and sodium hydroxides), volatile alkali (ammonia), and absorbent earth at the top of the columns for marine (hydrochloric) acid and nitrous (in modern terminology nitric) acid. Others had made similar insertions; but Geoffroy had

evidently thought them unnecessary, because those substances were covered by the column for 'acid spirits' in general. He included them in the column for vitriolic (sulphuric) acid merely to show that the 'oily principle or sulphur principle' came above them.

2. Tessari has added gold and platinum to the bottom of the column for nitrous (i.e. nitric) acid, separated by one space from silver just above them, as is gold in the column for marine (hydrochloric) acid in Geoffroy's table, and representing the fact that they are soluble in *aqua regia*. He has also added platinum to gold in the column for marine acid, perhaps following Lewis.

3. He has added zinc just above iron in the column for vitriolic acid.

4. He has added 'acid of vinegar' (acetic acid) to the columns for absorbent earth and volatile alkali (ammonia), just like Rouelle.

5. He has added arsenic at the top of the column for fixed alkali, and oil at the bottom of it.

6. He has added tin between lead and copper in the column for mercury, and platinum at the bottom of it.

7. He has added mineral acid and fixed alkali (in the same square) to the top of the column for water.

8. Like several others, he has added a column for spirit of wine (alcohol), showing its affinity for water and oil in that order.

The *table des Rapports* of B. G. Sage (1740–1824) omitted all the columns with individual metals at their heads except mercury, but included for the first time a column for phosphoric acid, and also like some others has columns for acid of vinegar (acetic acid), liver of sulphur, (potassium polysulphides) and phlogiston.[77] Phosphoric acid had been prepared by Margraf in 1743, thirty years before, and so was not a great novelty, though Sage himself had discovered hypophosphoric acid. However, there is an unexpressed assumption that affinity tables cover only the better-established substances. In a time when claims to have identified a new substance were quite common, and often shown to be unjustified, the reception of such substances into the canon was not necessarily rapid.

There are more substances in each of Sage's columns than in most previous tables, but those at the bottom of some of the columns are separated from those above them (like gold in the column for marine acid in Geoffroy's table) by blank spaces, presumably to show that their affinity is markedly weaker. In the column for mercury, the symbol for mercury is repeated at the start of a sequence lower down the column which is separated in that way from lead by two blank spaces; but this is probably a mistake for platinum, for which Sage has a very similar symbol. At the bottom of the table are shown some substances which do not unite with those at the head of the column, a device which resembles Gellert's section for substances which are insoluble and Rüdiger's rather similar section. Sage explains in his paper describing the table that the *rapports* of the substances depend on their specific gravity, quoting Buffon to support his identification of chemical affinity with gravitation.[78] His evident preference for the word *rapport* over *affinité* perhaps agrees with this view. *Affinité*, thanks to Demachy, had possibly recaptured for a time some flavour of the old idea that tendency to combine was due to similarity. Sage was later professor of assaying at the Paris Mint, and then the first Director of the École des Mines; and most of the experimental evidence which he quotes to support his table is mineralogical.

No copy seems to have survived of the table compiled by De Fourcy, the advocate of Meyer's *acidum pingue* (fatty acid), whose attack on the concept of affinity has already been quoted. However, the paper is described in Rozier's *Observations*.[79] He there says that although Geoffroy's table is no doubt the best imaginable way of showing the closest and most remote affinities of a number of substances, and though other tables have since been published by Grosse, Rüdiger, Gellert, and Demachy, he does not propose to examine the merits of these tables in detail. He considers that recent chemists have relied too much on analysis for knowledge of the composition of chemical substances. That is not enough, as some constituents are often destroyed in analysis, and synthesis must also be used. De Fourcy makes considerable use of the theory of phlogiston to explain reactions, and even more of *acidum pingue*. He describes his own table in the following terms:

> The table of chemical affinities which I here present has the advantage of naming the products. It comes a little nearer to synthesis. But it does not exempt us from looking for the reason for which one substance combines with another. It will be seen that it is always to *acidum pingue* that we must have recourse. Whether it is found in one substance or the other, or in both, or whether it is introduced by some preliminary operation, as has been said previously, we shall more readily provide a reason by this intermediary of the causes which combine two substances together than by the words affinity or relationships.
>
> This table is composed of thirty-six columns, at the head of which is placed the substance & all the substances with which it can combine. We have not been afraid of repetitions: they have even become necessary to give all the products of a substance without interruption. For example, in the first column mercury is placed at the head with all its combinations and products; but that does not prevent us, in the columns which have for the first substance acids or metals, &c., from again finding mercury and its products, and similarly for the rest.[80]

From this description it appears that the real novelty in De Fourcy's table was that he included the names of the products of the reactions, partly to show the composition of a compound ought to be judged from its synthesis rather than its analysis. Nevertheless, it is surprising that his paper is entitled 'Observations sur le tableau du produit des affinités chymiques', implying that the cause of the combinations which it represents is affinity. From what he has said about the true cause of chemical combination's being the presence of *acidum pingue*, although it is not at all clear how he supposes that acid may cause combination, it appears that he does not mean in using the word 'affinity' to imply that any similarity between the substances combining is the cause. In spite of Demachy, De Fourcy is clearly using the word 'affinity' again to mean no more than 'tendency to combine', with no deeper connotation at all.

His description of his table does not suggest that it would do more than its predecessors to distinguish the various complexities which might interfere with the workings of affinity, even if affinities were regarded as constant.

There are comparatively few differences between the table produced by J. C. P. Erxleben (1744–77) of Göttingen and that of Gellert published twenty-four years earlier.[81] Erxleben has put the table into words, and printed it continuously instead of in tabular form, with each series headed 'Step ladder of affinity' (*Stufenleiter der Verwandtschaft*). He is therefore claiming to show affinity and not merely solubility. He has also added columns for the 'inflammable substance', water, alcohol, 'aethereal oils', 'siliceous earth in general', acid of tartar (tartaric acid), and fixed air (carbon dioxide), and has added a number of substances to Gellert's lists without altering the order which Gellert had prescribed, except in one instance. In the series for the affinity of nitric acid he has replaced Gellert's cobalt by arsenic, with a question mark after it. This is hard to explain, since arsenic occurs higher up the list in both authors; but Erxleben is apparently trying to include the results of two reactions which show contradictory affinities.

He has kept Gellert's short list of exceptions, but has introduced them by the phrase 'do not dissolve', as if he were still giving a list of solubilities. In some places he has indicated his disagreement with Gellert's inclusion of substances among the exceptions, or added other notes of disagreement, as in the series for common salt; but he has not ventured to alter the series.

Bergman

As we have seen, the separation of affinities or attractions in the dry way from those in the wet way was foreshadowed in the tables of Geoffroy, Clausier, and Gellert, and more distinctly in that of Black. It is recommended by Baumé in his *Manuel de chymie* of 1763:

M. Geller [*sic*] has augmented this table considerably; but I think that it would be appropriate to make two tables of relationships at the same time. One would indicate the order of substances in the wet way, and the second the same order of affinities in the dry way. There is in Chemistry an infinity of instances which I shall point out in which substances which have no affinity at all in the wet way have it in the dry way, and vice versa. It is that which makes me think that this double table of affinities which I propose would be extremely useful.[82]

Professor Partington has also pointed out that Stahl had long before described a reaction which went differently in the dry way and the wet way.[83] However, the first published table to show explicitly the change in the order of affinity due to heat—or to claim to do so—was that of Torbern Bergman (1735–84) of Uppsala, in which the orders of affinities, or rather attractions, in the wet and dry ways are shown separately, as Baumé had suggested. The table was first published in 1775, but the much enlarged and revised version published in 1783 was the one which became accepted as standard and largely replaced Geoffroy's table.[84] Together with the long dissertation which introduces it, it was translated into French, German, and English within a few years of its appearance. Several English versions included slight alterations, and although these have been inserted in Table I (nos. 17b to 17h) they will not be separately discussed here.

After defining attraction and classifying different kinds of attraction, Bergman considers the fundamental question 'whether the order of attraction be constant'. He explains that there was a controversy on that point, 'one party contending, that affinities are governed by fixed laws, and the other affirming, that they are vague, and to be ascribed to circumstances alone'.[85] It will be noticed that although Bergman prefers the term 'attraction' he treats 'affinity' virtually as a synonym.

The latter party in the controversy was not in fact very strong: the critics of affinity tables more often argued that affinity acted differently in different circumstances than that there was no such thing as an underlying absolute order of affinity, and that the effects attributed to it were due merely to the combination of different circumstances, and no more. Bergman asserts that the controversy was only to be answered by consulting 'Experiment, the oracle of nature'.[86]

He does not, of course, approve of 'those general rules which affirm, that earths and metals are in all cases precipitated by alkalis, and metals by earths, for they are often fallacious'. Nevertheless, he believes that for each particular substance the order of attractions is constant, and that all apparent difficulties in accepting its constancy can be explained away. The order found in the wet way is the true order, and the differences found when the temperature rises above a certain point are simply variants. Fundamentally this is a matter of faith on Bergman's part, rather than of reason. He adds

But should there occur in this, as in other branches of natural philosophy, a few phaenomena, which appear to deviate from the ordinary track, they should be considered as comets, of which the orbits cannot yet be determined, because they have not yet been sufficiently observed. Repeated observations, and proper experiments, will in time dispel the darkness.[87]

Bergman attributes the effect of heat on the operation of chemical attraction entirely to the great differences in volatility between different substances at the same temperature. His argument is as follows. Let a substance A be attracted by two other substances; let the more powerful act at the ordinary temperature with force a, the less powerful with force b; and let the former be more volatile, its effort to rise being expressed by V, and that of the other by v. When these three substances are mixed together, the stronger will attract A with a force $(a - b)$; but if the temperature is gradually raised, V will increase faster than v until $a - b = V - v$. This equilibrium will be disturbed by the slightest increase in temperature, for then b, which was previously weaker will appear to prevail.[88]

Bergman here noticeably avoids talking about unobservable mechanisms, such as the then common suggestion (which he uses himself elsewhere) that the force of attraction between the particles of a substance was overcome by the force of repulsion between the particles of the matter of heat which surrounded them. Nor does he talk in merely general terms. Instead he uses the most precise and powerful method which he knew to describe the way in which the differences in volatility might operate—that is, he uses algebraic symbols. Although Newtonians such as Keill and Freind had used such symbols long before, there are few instances of practical chemists using them before Bergman. Nevertheless, the quantities to which he refers were not actually measurable in his own time, so that his suggested explanation could not be tested quantitatively then and there. They are only potentially measurable.

Bergman then discusses the various reasons which may produce apparent exceptions to the constant order of affinities in the wet way. The first type is 'Apparent irregularities from double attraction'. Such irregularities occur when double decompositions, caused by the sum of the attractions of one pair of substances being greater than the sum of the attractions of the other pair, are mistaken for cases of reciprocal decomposition involving only three substances. Very often the substance which has been left out of account is phlogiston. For example, the precipitation of metals dissolved in acids by other metals is never the effect of a single attraction, for during the solution a quantity of phlogiston is separated from the metals. (It is for this reason that in the later version of his table Bergman showed for wet reactions the calces (i.e. the oxides or in some cases other stable solid compounds) of the metals, which he believed to be elementary, rather than the metals themselves, which he believed to be compounds of phlogiston and the calx.) It will be recalled that Gellert had shown the calces rather than the metals in his column for reactions with glass (which would be in the dry way), but presumably not for the same reason.

Another example is the distillation of butter of antimony (antimony trichloride) from a mixture of corrosive sublimate (mercuric chloride) and regulus of antimony (metallic antimony). Neither mercury nor regulus of antimony will dissolve in marine acid (hydrochloric acid), unless a certain proportion of their phlogiston has first been removed. The process may therefore be explained by a double attraction: the calx (oxide) of mercury in the corrosive sublimate is attracted by the phlogiston, which is removed from the regulus of antimony, and the latter is then attracted by the marine acid.[89]

As an example of 'Apparent exceptions from successive change of substances' Bergman quotes Margraf's observation that although nitric acid separates marine acid from alkalis, the latter also dislodges the former. The explanation is that nitric acid expels marine acid by a single elective attraction; but the marine acid when poured on nitre (potassium nitrate) loses some of its phlogiston, because of the heat applied, to the nitric acid, which has a greater attraction for it. The nitric acid in this state is then expelled by that part of the marine acid which is not yet decomposed.[90]

The next class of irregularities is 'Apparent exceptions from solubility'. It may not be apparent, for instance, that a substance has been expelled from combination if it remains in solution and is not precipitated. This is a significant observation, because it implies that Bergman distinguishes between solution and genuine chemical combination. Many earlier compilers of tables had used solubility as a test for chemical affinity; but to Bergman it is not enough, as his description further on of his method for establishing elective attractions shows.

An example of this kind of apparent decomposition occurs when vegetable alkali (potassium hydroxide) is added to a solution of mineral alkali (sodium hydroxide) which has united with acids to the point of saturation (i.e. what is now called neutralization). The mineral alkali is not precipitated, because it is soluble; but if the solution is evaporated the vegetable alkali is found to be combined with the acid and the mineral alkali is uncombined. Sometimes the substance expelled from the combination itself dissolves the new combination, or at least does not hinder the water of solution from doing so. For instance, vitriolic (sulphuric) acid takes magnesia from the marine (hydrochloric) acid; but this is not apparent because water together with the released marine acid dissolves the vitriolated magnesia (magnesium sulphate), which only appears on evaporation.[91]

Among 'anomalous phenomena depending on apparent solubility' Bergman mentions that siliceous earth (flint) appears not to be precipitated on pouring liquor of flints (flint dissolved in a solution of fixed alkali, i.e. potassium or sodium hydroxide) into acid if the liquor is diluted with twenty-four times its own weight of water or more. This is due not to solution but to siliceous particles' being so dispersed in the abundance of water that they cannot subside on account of the great proportion of their surface to their weight.[92]

'Exceptions from the combination of three substances' occur chiefly in the dry way, with earths and most metals. For instance volatile alkali (ammonia), marine (hydrochloric) acid, and the calx (oxide) of mercury combine, and so do volatile alkali, vitriolic (sulphuric) acid, and magnesia, and iron, vitriolic acid, and magnesia. Compounds of four ingredients are also formed, such as those of borax (sodium tetraborate) with tartar (potassium hydrogen tartrate), vitriolated magnesia (magnesium sulphate) with common salt, gypsum (calcium sulphate) with common salt, and many others.[93]

The last type of apparent exceptions which Bergman discusses is 'Exceptions from a determinate excess of one or more of the ingredients'. Many examples show that a determinate excess of acid can be accepted by neutral salts.[94] (That is, in modern terms, they may be converted to acid salts, as Rouelle had demonstrated long before Bergman's time.[95]) For instance, if to a saturated solution of neutral tartarized tartar (potasium tartrate) in distilled water some genuine acid of tartar (tartaric acid) is added, real tartar (potassium hydrogen tartrate) is precipitated. The tartar is simply an acid salt, and is also separated by the addition of any other acid. This is usually explained by saying that the acid expels the tartar by a superior attractive force. However, that does not explain why the alkali of the tartar is precipitated along with the acid, nor why the precipitation should take place on the addition of either the acid of tartar itself or the acid of vinegar (acetic acid), which is a weaker acid.

That we may distinctly perceive what happens in this operation, let the tartarized tartar be imagined to be divided into two parts, so that one part *b* shall contain as much acid as is necessary for the other *a* to become tartar. Now let the foreign acid be added, so as to saturate the alkaline basis of the part *a*, the acid of tartar before combined with it will flow back to the portion *a*, which already tends to it with so much force, that it immediately seizes it, and is converted into tartar, provided anything capable of weakening the cohesion of the principles in *b* but in a small degree be added.[96]

Bergman gives a similar interpretation of Baumé's decomposition of vitriolated tartar (potassium sulphate) by nitric acid in the wet way. That implied that nitric acid had a stronger attraction for the alkali than had vitriolic (sulphuric) acid, although on other grounds it was supposed that vitriolic acid had the stronger attraction. Baumé's reaction therefore appeared to be an example of reciprocal affinity. Bergman, however, makes the following observations. (1) Vitriolated tartar dissolved in water may be crystallized by evaporation, after the addition of a quantity of vitriolic acid equal to one-third of the salt. The crystals, with the accession of one-third of their weight, remain dry, in spite of being acid. More acid produces a deliquescent salt. The excess of acid cannot be easily driven off. (2) Vitriolic acid in the proper quantity completely decomposes nitre (potassium nitrate) even in the moist way, which confirms that its powers of attraction are stronger. No distinction therefore need to be made between the dry and wet ways. (3) If vitriolated tartar is dissolved in hot, strong nitric acid, only a third of it or a little more is decomposed, however much acid is used. (4) This much decomposition will take place even if the nitric acid is cold and dilute. (5) Vitriolated tartar combined with a proper excess of acid, as in Observation 1 above, is not changed at all even by the most concentrated nitric acid. (6) Not only the nitric acid but several others decompose vitriolated tartar. (7) The two-thirds of the vitriolated tartar which remain unchanged form crystals with the vitriolic acid released from the other third in the way described in Observation 1.[97]

Bergman then explains these observations in the following way. (The eighteenth-century translator uses the term 'nitrous acid' for what we now call 'nitric acid', the distinction between the two having not yet been made clear.)

Suppose *b* to be such a portion of the vitriolated tartar, as to contain exactly that excess, which the other portion *a* can receive. Nitrous acid of itself cannot deprive the vitriolic of its basis; but *a* attracting it at the same time, so far diminishes the resistance, that the nitrous is able to seize the alkaline basis of *b*, but its power is confined to certain limits. Suppose the vitriolated tartar to be divided into two parts, one of which affords its basis to the nitrous acid, and the other is not decomposed. We have here three powers: let that by which the part of the vitriolated tartar remaining entire attracts a determinate excess of acid, be called *A*; *B*, that by which the part to be decomposed endeavours to retain its basis; and, lastly, *C* the force of attraction of the nitrous acid to the same basis, it is obvious that no decomposition can be effected, if $A + C < B$, or if $A + C = B$; but if $A + C > B$, it immediately takes place.[98]

Elsewhere, Bergman, unlike Berthollet twenty years later, does distinguish between the excess of a substance which may be needed to make a reaction go to completion, and the quantity of the substance which is actually included in the resulting compound.[99] To a modern chemist that would no doubt seem an obvious and essential point; but to an eighteenth-century chemist, whose understanding of the composition of the compounds was supported only by a vague general concept that matter consisted of particles and not by a quantitative atomic or molecular theory, it was by no means an obvious point. The thoroughness with which Bergman applies his model of the nature of reactions in solution, using the results of his experiments honestly with a minimum of preconceptions, makes his work an important contribution to the progress of chemical theory, if only because he made the true problems clearer. However, there are in his thinking certain conceptual barriers which were hardly overcome before the work of Guldberg and Waage. These barriers will be discussed in the next chapter.

Having enumerated the various types of exceptions or apparent exceptions to the constancy of affinities, Bergman goes on to explain the methods by which he established his table. For single elective attractions, he tested whether the addition of a substance *c*, say, to a solution of a compound *Ad* precipitated *Ac*. If not, the solution was evaporated to dryness and the composition of the resulting crystals was determined. The fallacy of supposing that because a substance dissolved or was not precipitated, it must therefore be held in chemical combination, was thus avoided. The process was repeated for the dry way in a crucible or retort. In existing tables, Bergman justly remarks, only a few substances were shown and compared with a few others.

The slight sketch now proposed will require above 30,000 exact experiments, before it can be brought to any degree of perfection. But when I reflected on the shortness of life, and the instability of health, I resolved to publish my observations, however defective, lest they should perish with my papers, and I shall relate them as briefly as possible. In itself it is of small consequence by whom science is enriched; whether the truths belonging to it are discovered by me or by another. Meanwhile, if God shall grant me life, health, and the necessary leisure, I will persevere in the task which I have begun.[100]

His health had been weak since he strained it through overwork as a student, and very poor since 1769. He retired from active work in 1780, three years before the publication of the revised version of the *Dissertation*, and died in the year after its publication. He was forty-nine years old. No earlier compiler of such a table seems to have carried out any such systematic series of experiments on which to base it. Most of the facts on which the tables were founded were already available, special experiments were done only for disputed cases, and the fallacy which Bergman mentions of supposing that a substance which was dissolved or not precipitated must be held in chemical combination was certainly often committed.

Most of the text of the *Dissertation* is made up of a detailed exposition of the table of single elective attractions. Since the table included most of the substances normally used in chemical laboratories at the time, this exposition is virtually a commentary on the whole of the known facts of inorganic chemistry, with a little organic chemistry (then hardly even in its infancy) as well. There are fifty columns in the original version of 1775, whereas the previous highest number of separate substances shown at the head of an affinity series was Erxleben's thirty-six. The 1775 version was printed on a single sheet, but the revised version of 1783 had fifty-nine columns and occupies two sheets. Columns have been added in it for acid of benzoin (benzoic acid), acid of amber (succinic acid), acid of sugar of milk (mucic acid), acid of fat (probably a mixture of stearic, oleic, and other fatty acids), *acidum perlatum* (phosphoric acid, though Bergman did not believe it to be such), acid of Prussian blue (prussic acid), matter of heat (the imponderable fluid thought to produce the phenomena of heat), and siderite (iron phosphide, though again Bergman did not believe it to be such). Scheele showed in 1785, the year after Bergman's death, that *acidum perlatum*, a component of urine, was the same as phosphoric acid, and siderite, which Bergman believed to be a distinct metal, was really iron phosphide.[101] The inclusion of a number of organic acids, several of them newly discovered by Scheele, is interesting as a mark of the very early stages of organic chemistry. Bergman treats them each as a unitary substance and does not consider their composition. In fact hardly anything was known about it, which is the reason why these substances were named simply according to their natural source. If the same acid could be derived from two different sources, it was easy to mistake the samples from the two sources for different substances.

In his table Bergman follows further than his contemporaries the logical consequences of regarding matter of heat and phlogiston as chemical substances. He explains his view of the nature of heat in discussing combustion in the 1783 edition. Regarding fire as the action of heat increased to a high enough level to be visible, he rejects the old conception of heat as the motion of particles of bodies, arguing that action left to itself quickly dies out, whereas a small spark will quickly become a large fire. He summarizes the three main opinions then current on the nature of fire:

1. That fire is light, and when it enters into chemical combination it becomes phlogiston. Although the most obvious facts of combustion would be explained by this theory, Bergman rejects it because phlogiston has been shown to be the same as inflammable air (i.e. what is now called hydrogen).

2. That fire is an element quite distinct from phlogiston, and expels it from chemical combination.

3. That heat is a compound of phlogiston and vital air (oxygen), a theory of Scheele's which Bergman supports.[102] It explains why vital air

disappears from the atmosphere when combustion takes place, as it is supposed to combine with the phlogiston to form heat. If the heat then passes through the walls of the vessel in which the burning substance is enclosed, the theory also explains why the volume of air remaining in the vessel contracts. Bergman also mentions Kirwan's theory that vital air and phlogiston combine to form what would now be called carbon dioxide, which would explain why organic substances give carbon dioxide when they burn.[103]

Scheele's theory, on the other hand, would explain the fact, to which Lavoisier had drawn attention, that some substances gain weight when they burn. For it could be supposed that they absorbed the heat, which, being composed of vital air and phlogiston, had weight.[104] After all, even in Lavoisier's theory heat was supposed to be material and to be emitted in burning. Elsewhere, on the theory that charcoal is a compound of phlogiston and aerial acid (his name for what is now called carbon dioxide), Bergman deduces the weight of phlogiston contained in it by subtracting the weight of the residue of ash and of the aerial acid produced from the combustion of a known weight of charcoal. By further weighing of the reagents and products of reactions he proceeds to estimate the weight of phlogiston in various metals.[105]

We must beware of hindsight and the danger of reading the attitudes of late-twentieth-century chemists back into the circumstances of the eighteenth century; yet Bergman's thoroughness in treating phlogiston as an ordinary chemical substance, with measurable weight and specific elective attractions, like any other substance, leaves the modern reader with a feeling that if he had lived he must surely have found some flaw in the phlogiston theory itself. It had survived for so long partly because the properties of phlogiston were so vague and adaptable that the theory could not be falsified. However, at this stage Lavoisier's theory was hardly more satisfactory, as Professor Schofield has pointed out.[106]

In the 1775 version of Bergman's table the metals themselves are shown at the heads of their columns; but in the revised version he has taken to its logical conclusion the belief that the reaction of a metal with the solution of another metal in an acid is a double decomposition. The metal is a compound of its calx with phlogiston, and the single elective attraction to be taken into account is that of the calx for the acid. It is therefore the calx and not the metal which should be shown as reacting with acids in the wet way.[107]

The order of attractions of one acid for the various metals is much the same as that of another, and so the twenty-five columns for acids are very much the same as each other. Indeed, some of Bergman's followers simplified his table by combining several columns together. In a sense, the similarity of the columns is the important thing, and points to a simple underlying pattern which is not obvious in all cases because so many other factors affect the result of a reaction. Bergman, however, is too austerely empirical to assume such a simple pattern, and shows each acid separately to bring out small variations in the order of the substances in each column. Nevertheless, as far as possible where several substances occur in the same order in successive columns he has put them on the same rows. That makes the similarity between columns clear even if blanks have to be left in some of them where the substance in that row in other columns does not happen to react with the substance at the head of the column, and so does not belong in it.

There are many series of symbols in Bergman's table that are not separated by horizontal lines. Evidently this means that the substance at the head of the column has equally strong attractions for all of them, or at least that Bergman has not yet established any difference.[108] Such bracketing of substances together is often found in earlier tables, but not so extensively. Within these groups little is conveyed of the information which is intended to be the table's chief function.

In the text of his dissertation Bergman suggests that the purpose of the separate section for attractions in the dry way is to show how the order of attractions is modified by heat.[109] However, in many cases the substances shown as combining in the dry way are by no means the same as those combining in the wet way. For instance, in the columns for metals (or their calces in the 1783 version) the substances shown as combining with them in the wet way are mostly acids, but in the dry way (naturally) they are mostly other metals. Hence the section for the dry way shows that the substances which combine in the dry way are different from those which combine in the wet way, and not that the order of attraction is different for the same substances.

Diagrams of double attractions

One of the points most commonly made against affinity tables was that in practice reactions very often involved more than three substances, whereas the tables could show only the relative affinities of two substances for a third. Bergman, however, was the first compiler of a published table to find a means of showing the result of reactions involving more than three substances. Before his table of single elective attractions he gives another table in which are sixty-four diagrams showing the results of reactions in which there are several reagents. As we have seen, in this Bergman had been anticipated by Black in the table accompanying his lectures.

The earliest known diagram of this kind was published by Jean Béguin in the early seventeenth century to show the reaction of corrosive sublimate (mercuric chloride) with the sulphide of antimony (Fig. 4.1).[110]

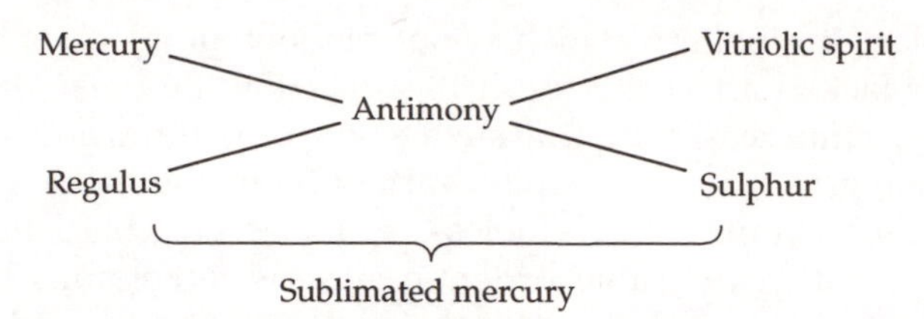

Fig. 4.1 Jean Béguin's diagram of the reaction of corrosive sublimate (mercuric chloride) with the sulphide of antimony. From Jean Béguin, *Elémens de chymie*, Paris, 1615, pp. 167-8.

However, the relationship between Béguin's diagram and those of the eighteenth century can be no more than a conjecture.

As has already been mentioned, such diagrams were used in his lectures at Edinburgh by William Cullen.[111] In these the substances involved are shown at the four corners of a rectangle, or three corners if only three substances are concerned. Pairs of substances which are in combination before the reaction are on the same side of the rectangle and are joined by a bracket. The result of the reaction is shown by joining the two substances between which the attraction was strongest with one or two arrows. Cullen explained how the relative attractions could be read from the table of attractions (in a MS quoted by Professor Crosland), referring oddly enough to Geoffroy's table and not one of his own.

Cullen also used another type of diagram, which suggested a mechanical analogy and the possibility of quantifying the forces involved.[112] The substances are again shown at the corners of a rectangle, but those at opposite corners are joined by lines which Cullen called 'rods'. The strength of the attractions is represented by letters written in the angles where the lines cross. The affinity table would show which pair of attractions was the stronger, and so the letters could be used to form an equation, or rather an inequality, showing how the reaction would go.

See for example Fig. 4.2.

Black is represented by Robison, the editor of his lectures on chemistry, as having used diagrams of this second kind (apparently after Cullen had introduced them);[113] but in one of the surviving manuscripts of notes of his lectures the relative strength of the attractions is represented by numbers instead of by letters.[114] However, these numbers are not chosen to represent the magnitudes of the forces absolutely, but are quite arbitrary, and chosen merely so as to show students how the observed result for the reaction would follow.

Nevertheless, Black eventually felt that such schemes implied mechanisms for which there was no evidence. In a later type of diagram he avoided implications of that kind. A horizontal line bisected two circles drawn side by side. In the four semicircles thus formed were written the symbols for the four substances taking part in a reaction, or for several

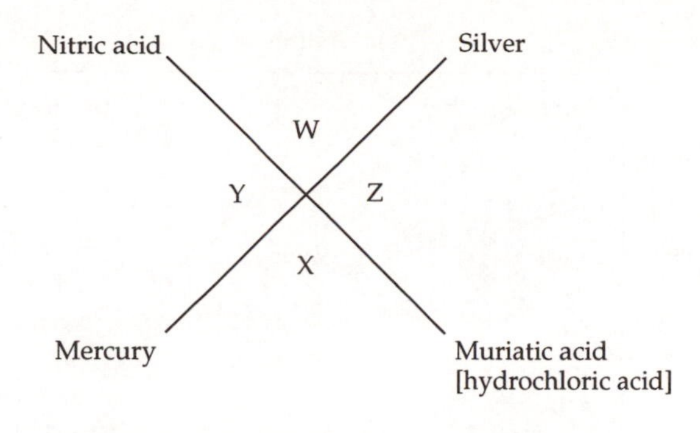

'Y > X and Z > W by table.
Ergo Y + Z > X + W.'

Fig. 4.2 Cullen's diagram of the decomposition of mercuric nitrate by silver chloride to form mercuric chloride and silver nitrate, with 'rods'connecting the substances and the strength of the attractions represented by letters written in the angles where the lines cross. By looking up the affinity table to see which pair of attractions was the stronger the letters could be used to form an inequality showing which way the reaction would go. As illustrated in M. P. Crosland, 'The use of diagrams as chemical "equations"', *Annals of Science*, **15**, 1959, pp. 79-80.

substances with similar properties.[115] This is the type of diagram in the MS of Black's lectures of about 1767/8 that has been discussed above (p. 129).

The lectures of Black and Cullen were well known in Europe through the reports of their pupils, and there can be little doubt that Bergman's diagrams are somehow derived from them, in spite of obvious differences. He carefully avoids anything that might suggest a mechanical explanation. The arrows in nos. 9–11 mean simply that one substance is partitioned between the two others, and do not mean that there is any force acting. This notion of partition is related to Bergman's appreciation of the importance of the relative quantities of substances in a reaction. Another innovation is the use of the symbol for water or heat in the middle of the diagram, showing that the reaction takes place in the wet way or the dry way. According to his views in 1783 Bergman should logically have counted heat as one of the substances taking part in the reaction, contributing to the result with its own attractions. However, he has not amended the diagrams of double elective attractions in that way, and it is clear that if the symbol for heat means anything more than that the reaction is carried out in the dry way, the heat is regarded as merely a medium in which the reaction takes place. The same applies to water for reactions in the wet way. The points of the horizontal brackets are turned upwards if the products remain in solution, or downwards if they are precipitated or otherwise displaced.

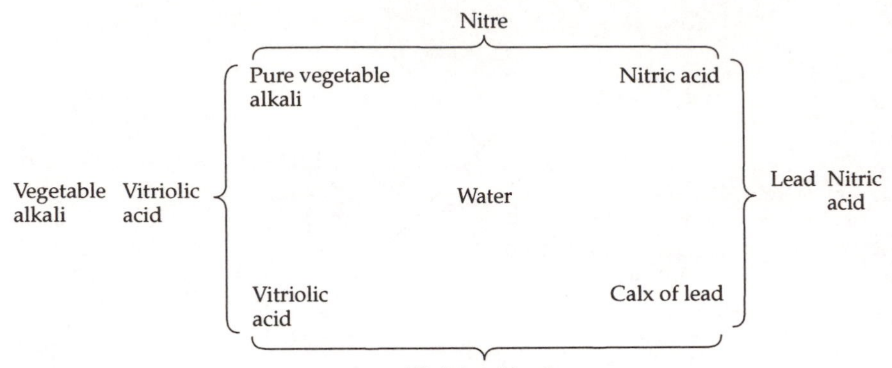

Fig. 4.3 Bergman's diagram No. 22 transcribed into words, showing how vitriolated vegetable alkali (potassium sulphate) and nitrated lead (lead nitrate) in solution react to produce nitre (potassium nitrate), which remains in solution, and vitriolated lead (lead sulphate), which is precipitated. From Tobern O. Bergman, *Dissertation on elective attractions*, London, 1970 (1785).

As an example of Bergman's diagrams, no. 22 (Fig. 4.3) is here transcribed into English words, which are much more cumbrous than the alchemical symbols.

Figure 4.3 means that vitriolated vegetable alkali (potassium sulphate) and nitrated lead (lead nitrate) in solution react to produce nitre, which remains in solution, and vitriolated lead (lead sulphate), which is precipitated.

Bergman's table of such diagrams of double decompositions is far more extensive than Black's,[116] and covers a large part of the inorganic chemistry then known. They represent a more determined attempt to reduce to system and pattern, which the century so much admired, one more aspect of the apparent disorder of phenomena. They became a more fruitful attempt than the tables of simple elective affinities or single elective attractions, for they are in the line of thought which led to chemical equations and to nineteenth-century attempts to represent chemical processes and the composition of molecules by diagrams.

Tables after Bergman

After Bergman's enormous efforts, and his confession that the necessary experimental work was not nearly complete, only an exceptionally hardy spirit could have faced the huge task of improving radically on the tables which Bergman had left. Only by a vast extension was it possible to produce an affinity table which would come closer to taking into account all the conditions which govern the course of chemical reactions. The attempt was made by Weigel in 1777, even before the publication of

Bergman's revised version.[117] Weigel practically abandoned the orthodox type of table. His lists of substances are in words, like Erxleben's, and the first section is headed merely 'Examples of Combining Affinities'. This section consists of two parts, the first for 'appropriative affinities'[118] and the second for 'predisposing affinities', each divided into a column for the wet way and a column for the dry way. In the part for appropriative affinities, pairs of substances on the same line are bracketed together, with a third substance in between them on the line below. The meaning is that the two substances on the upper line combine when brought together by the substance on the lower line, which exercises the 'appropriative affinity'. In the part for 'predisposing affinities', sets of three substances are arranged in the same way, but the meaning is that the substance on the lower line predisposes the right-hand substance to unite with the left-hand substance.

The next section is headed 'Examples of Decomposing Affinities with a Single Combination', and is divided into a part for the wet way and a part for the dry way. Each comprises a number of lists which are very much like the columns of a normal affinity table, except that, as in the tables of Gellert, Spielmann, and Erxleben, it is the lower substance which displaces the higher from combination with that at the top of the list—in other words, the usual order is reversed. The third section is headed 'Examples of Decomposing Affinities with a Double Combination' and consists of diagrams like Bergman's diagrams of double decompositions. The last section consists of diagrams set out in much the same way but with more than four components involved, and is headed 'Examples of Decomposing Affinities with More Combinations'. To sum up, Weigel has gone to some lengths to make his table show the complex ways in which three, four, or more substances interact, though he has not managed to allow for various other complicating factors such as variations in concentration. However, in allowing for as wide a range of differing situations as he does, he has thoroughly obscured any possibility of seeing regular patterns which might lead to general laws of chemical combination, or indeed even regular patterns which might help students to grasp and remember the observed tendencies of substances to combine.

Wiegleb's table is obviously derived from Weigel's in form.[119] He has added an introductory section, which is headed 'Examples of the Affinities of the Basic Mixt, Which are Accompanied by a Combining of Substances, and also through Which an Altered Basic Mixt is Produced'. This simply gives examples of kinds of substances which combine together, or one of which dissolves the other, in cases where neither is already combined with another substance, so that there is no question of displacement. This section is divided into a part for simple combinations of two substances and a part for multiple combinations of more than two substances. Wiegleb tends in these first two sections to include substances which would now be regarded as mixtures of indefinite composition, and one of his two examples of 'affinities of the basic mixt' is that of oils with 'tallow, wax, lard and butter'. At this date, as we have seen, there was still very little understanding of the composition of organic compounds.

His second section gives examples of combining affinity, and has two parts headed 'complete' and 'incomplete', corresponding with Weigel's 'appropriative' and 'predisposing' affinities. Both parts have divisions for the dry way and the wet way. Wiegleb writes the three substances concerned on the same line, with the mediating substance bracketed between the other two. His third section is headed 'Examples of Decomposing Affinity of the Basic Mixt' and is similar in arrangement to Weigel's second section, although the substances included are considerably different. Wiegleb's last two sections, for double decompositions and multiple decompositions respectively, are modelled on Weigel's last two sections, but have considerable differences in the contents.

Both these tables have obviously derived from Bergman the distinction between the wet way and the dry way and the diagrams of double decompositions; but they also owe something to the tradition of Gellert and Erxleben, and contain a good deal of material which had not previously been included in a table at all. In the attempt to allow for so many complexities Wiegleb, like Weigel, has gone to the extreme of showing almost every reaction separately, and so quite lost sight of the possibility of showing any general pattern. He has reached a dead end.

Several slight variations on Bergman's table were published in Britain after Beddoes's translation of the *Dissertation on elective attractions* appeared in 1785. However, in 1782 John Elliot (1747–87), a believer in the negative weight of phlogiston (which he called 'electron'), had already inserted a slightly altered version of Bergman's table in his *Elements of the branches of natural philosophy connected with medicine.*[120] Elliot found that affinities varied according to the solvent. For instance in water a soap of olive oil and slaked lime (calcium hydroxide) was not decomposed by mild fixed alkali (sodium carbonate), but in spirit of wine (alcohol) it was.[121] Elliot also inserted in his own book a much shorter Table of Elective Attractions of his own.[122] It is intended to give the reader a preliminary view of how the full table worked rather than to be complete. There is only a single column for all metals, containing the four mineral acids and 'acetous' (acetic) 'acid' and nothing else; and although like Bergman's table it includes columns for fixed air (carbon dioxide) and 'pure air' (oxygen), it omits nearly all Bergman's columns for organic acids.

The massive table produced by Gergens and Hochheimer in 1790 is directly derived from Bergman's table of 1783.[123] They have no column for phlogisticated vitriolic or nitric acids (sulphur dioxide and oxides of nitrogen respectively), dephlogisticated marine acid (i.e. chlorine), *aqua regia*, the acids of sugar of milk or of milk (mucic and lactic acids), acid of Prussian blue (prussic acid), *acidum perlatum* (phosphoric acid under another name), or siderite (iron phosphide under another name). (The last two were not, of course, distinct substances, as Bergman had supposed.) They combine the three alkalis in one column, add a column for 'acid of wood vinegar' (probably acetic acid distilled from wood), and replace 'matter of heat' by 'fire', recalling Clausier. Also, they name the products of the reactions between the substance at the head of the column and many of the substances within the column, as De Fourcy had apparently done, and more recently Weigel. Apart from the fact that Gergens and Hochheimer's table is in words and not in symbols, other differences from Bergman's are few. Instead of showing the order for attractions in the dry way underneath, they show a separate column for the dry way alongside that for the wet way in twenty-six of their fifty-nine columns, and add a note showing the few differences in the dry way for eleven of the acids, for which there is no second column showing the attractions for the dry way, from the two acids for which there is such a column. They vary slightly the order of the organic acids so as to separate the vegetable acids from the animal acids. Otherwise they follow the order of the columns of Bergman's 1783 table almost exactly.

The table in the first edition of the *Systematisches Handbuch der gesammten Chemie* of F. A. C. Gren (1760–98) is fairly conventional.[124] It has fifty-three columns, with words instead of symbols, and distinguishes between the wet way and the dry way. It is very much like Bergman's table, but Gren has included some substances, such as resin and gum, which the more discriminating Bergman had avoided, presumably because he appreciated that they were not simple substances. However, in the second edition of the same book Gren adopted his own version of the oxygen theory, which was intended to fit it into the framework of the phlogiston theory. In this edition he has a table in quite a different form, resembling those of Weigel and Wiegleb more than Bergman's. It is not in the form of a table with columns, but is made up of separate sections for the affinities of eighty-seven substances, with numerous examples of double decomposition. The first section, occupying twelve pages, shows the chemistry of oxygen according to his theory. An example of one of his diagrams will illustrate his ideas. It refers to a reaction at a temperature of over 500 °F (Fig. 4.4).

This means that oxygen gas (*Sauerstoffgas*), which as in Lavoisier's theory consists of oxygen and matter of heat (*Wärmestoff*—the equivalent of Lavoisier's 'caloric') combines with sulphur, which consists of 'sulphureous matter' (*Schwefelstoff*) and 'inflammable matter' (*Brennstoff*, or phlogiston). The products are the acid of sulphur (*Schwefelsaures*),

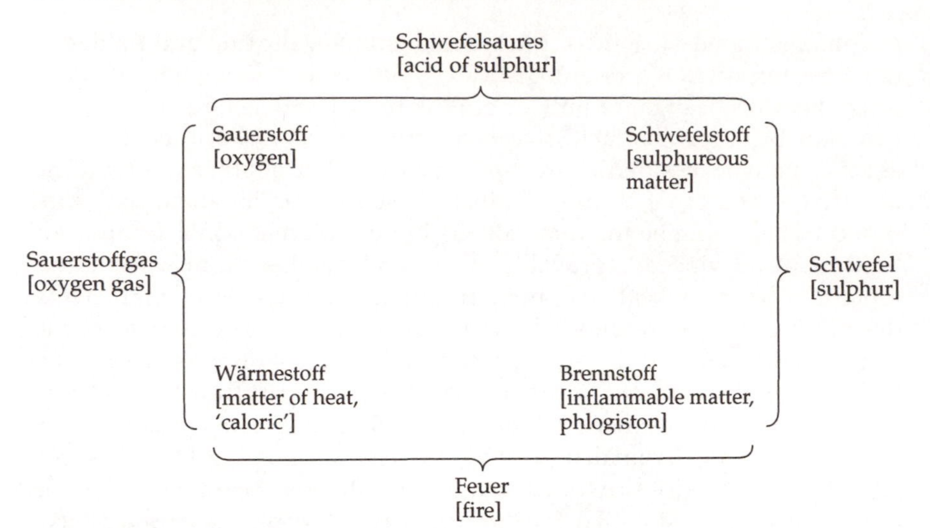

Fig. 4.4 F. A. C. Gren's diagram for the second edition of his *Systematisches Handbuch der gesammten chemie*, 4 vols, Halle, 1794-6, Vol. IV (1796), pp. 143-216, showing that at a temperature of over 500 °F oxygen gas (*Sauerstoffgas*), consisting of oxygen (*Sauerstoff*) and ' matter of heat'(*Wármestoff*), combines with sulphur (*Schwefel*), consisting of ' sulphureous matter'(*Schwefelstoff*) and ' inflammable matter'(*Brennstoff*, or phlogiston), to produce acid of sulphur (*Schwefelsaures*), formed from the sulphureous matter and the oxygen, and fire, formed from the matter of heat and inflammable matter (i.e. from Lavoisier's ' caloric'and phlogiston). Like Lavoisier and other contemporaries, Gren regarded what would now be called the acidic oxide of sulphur as the acid proper, any water in which it might be or become dissolved being regarded as an insignificant incidental rather than as producing a significantly new compound.

formed from the sulphureous matter and oxygen,and fire, which is a compound of the matter of heat and the inflammable matter, or in other words caloric and phlogiston. Like Lavoisier and others of his time, Gren regarded what we would now call the acidic oxide of sulphur as being the acid proper, the water in which it might be dissolved being regarded as an incidental.

The whole of inorganic, and some of organic, chemistry is interpreted most ingeniously in this manner. The table is a clear presentation of Gren's theory, but it is so long-winded, with a separate diagram for every reaction, that it does not serve any of the usual purposes of affinity tables. The much simpler orthodox table in the abridged version of the book is much easier to follow, though it is not tabulated except in the English translation.

J. B. Trommsdorff (1770–1837) of Erfurt published a set of twelve tables, of which eight show the names of the products of the combination of the various elements and radicals, and four show affinities.[125] Among the four showing affinities, the first has six columns for alkalis (including 'baryt' (barytes, i.e. barium sulphate) and 'strontian' (strontium carbonate)), five for earths, and sixteen diagrams illustrating examples of double decomposition. The next has thirty columns for the simple affinities of acids with alkalis, earths, and metallic oxides in the wet way, with forty-five examples of double affinities at the bottom; the third has seventeen columns for the simple affinities of metals with acids in solution at ordinary temperatures, again with forty-five examples of double affinities at the bottom; and the fourth, corresponding with Bergman's table of diagrams of double decompositions, has eighty examples of the double affinities of decomposition of metallic salts. The table is clear and well organized, and shows substances according to the normal oxygen theory of the day. Although it owes something to Weigel and Wiegleb, it is more like Bergman's. In most respects it was the best of all the affinity tables; but their day was almost past by 1800.

General tendencies

It is difficult to draw up a simple family tree showing the line of descent of affinity tables, as each compiler of a table evidently studied the work of several of his predecessors, took the features which he liked from each, and added a few new points of his own. It is therefore impossible, except where only a few modifications have been made to a previous table, to show that any table is wholly derived from a single predecessor; and even marked similarities may have been conceived independently.

Nevertheless, it is clear that there are two main traditions. All tables are descended from Geoffroy's, which is the origin of the first main tradition; but several, especially of the German ones, stand in a rather different tradition, and were influenced by Gellert's distinctive innovations. These are combined by Bergman with his division between the wet and dry ways and his introduction of a range of organic acids into a pattern which was predominant until the end of the century. This pattern was associated with the diagrams of double decompositions which Bergman had probably derived from the Scottish school. The tradition represented by Bergman and Gellert is more concerned with metallurgy than the French tradition, which is more concerned with theoretical and methodological objections to the tables. Figure 4.5 is a rough indication of what few affiliations can be tentatively traced between the various tables.

The tables also show differing ways of trying to correct the deficiencies of their predecessors. One way is simply to enlarge the table by including more and more substances, though no great significance need be attached to the fact that a graph of the maximum number of substances at the heads of columns is roughly an exponential curve, with a doubling time of about thirty years. This increase in size is partly due to the

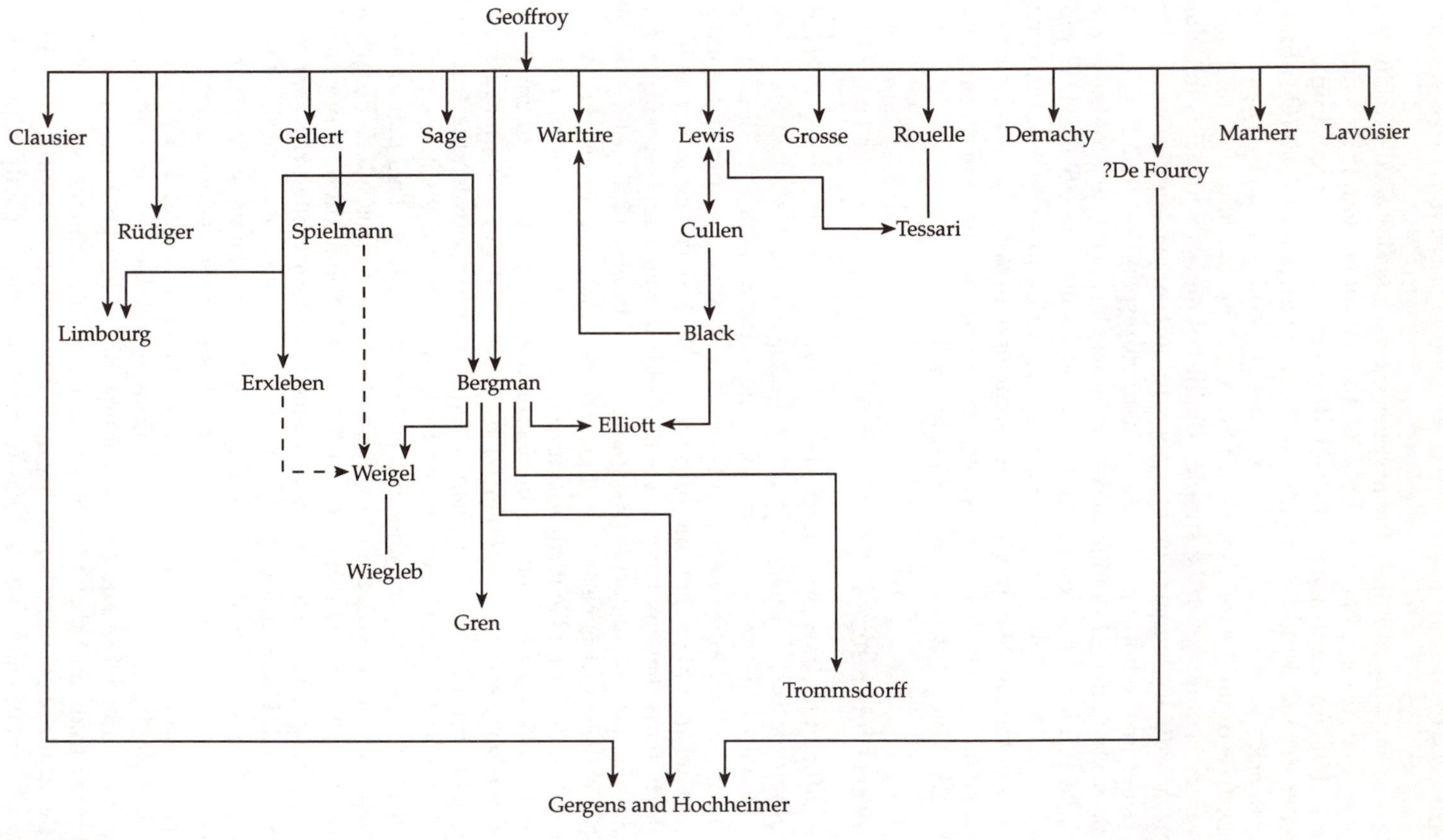
Geoffroy
Clausier
Gellert
Sage
Warltire
Lewis
Grosse
Rouelle
Demachy
?De Fourcy
Marherr
Lavoisier
Rüdiger
Spielmann
Cullen
Tessari
Limbourg
Black
Erxleben
Bergman
Elliott
Weigel
Wiegleb
Gren
Trommsdorff
Gergens and Hochheimer

Figure 5

increase in the number of substances known or in normal use; but it is also due to a wish to show different substances separately, and not to imply that a number of substances as a class followed the same order of affinities.

Opposed to this tendency for growth in size, however, is a tendency to simplify, because the larger tables implied many reactions which were not fully understood or were even the reverse of what could be observed experimentally, and also no doubt because of the French or British desire to show the main lines clearly and not to make the tables too elaborate. In addition, there is the repeated tendency to show exceptions, or variations, or distinctions between affinities under different conditions. However, the orthodox type of table inevitably suggested that one order of affinities was the normal one, and that any contrary instances were merely variations; so that it was not until Weigel and Wiegleb practically abandoned the orthodox type that different circumstances could be shown on an equal footing—and under their enormous sledge-hammers the table became little more than a list showing every reaction separately, and losing the advantage of tabulation.

Lavoisier, in a paper read to the *Académie* in 1783, presented a column showing the affinities of his *Principe oxygine* (oxygen), making clear the exceptions due to different degrees of heat, to be added to existing tables of the conventional form.[126] However, in the same paper he summarized the known difficulties in using that form:

1. The tables present simple affinities—but we find in Nature only double, triple, or still more complicated affinities. (This is the difficulty which Bergman in his diagrams of double decomposition, and Weigel and Wiegleb in their even more elaborate diagrams, tried to overcome.)

2. They take no account of the 'fiery fluid, principle of heat', which is all-pervasive. The resistance of the constituent molecules of bodies to separation is only the resultant of two variable forces: (i.) obedience to a certain law relative to the degree of the thermometer; and (ii.) corresponding to the greater or less separation between the particles of the body caused by the introduction of the 'fiery fluid'. Hence the same body can be solid, liquid, or aeriform according to the vigour with which the matter of fire tries to separate the particles. Another factor is atmospheric pressure. Consequently the effect of combining two bodies differs with temperature, and a Table of Affinity can present the true results at only one degree of heat. The changes with temperature in the state of combination of mercury and oxygen are an example. Bergman gives two tables: there should really be one for each degree of the thermometer. (In fairness to Bergman, it should be remembered that in the 1783 version of his table, which Lavoisier presumably had not seen when writing this paper, dated 1782, there is a column for affinities with matter of heat.

Hence reactions where heat was concerned could be treated as cases of double or more complex attractions. However, Bergman does not take that line of argument very far.)

3. The tables do not take account of the attraction of water in considering reactions in the wet way.

4. They do not express variations in the attractive force of *molécules* (particles) of bodies due to different degrees of saturation, for instance in sulphurous and sulphuric acids (then fairly recently distinguished). The affinities of simple substances should be given, in this instance the affinity of sulphur without oxygen. Lavoisier does not in fact follow this advice in his *Traité elémentaire*.

The intended functions of affinity tables

The development of affinity tables in the eighteenth century provides a clear insight into the evolution of chemistry as an autonomous discipline with its own distinctive problems, concepts, and preoccupations. In particular, it gives us insight into the growth of a distinctively chemical view of what kinds of matter should be reckoned as principles or elements or simple substances, of which they constitute a kind of unofficial list. However, they served in practice a number of other purposes, among which perhaps the most obvious is that they might help the novice, or someone who needed to use chemistry without fully understanding it, to remember the facts which they summarized. Handbooks of pharmaceutical preparations for use in dispensaries frequently included one or other of the standard affinity tables during much of the eighteenth century, and indeed even in the early nineteenth century.[127] Lewis's *New dispensatory* is an example, rather more philosophical than most.

However, another intended function of affinity tables was that they might lead the experienced chemist to predict the result of a reaction which he had not yet tried. Nevertheless, it is hard to believe that any of the tables were often used in this way, or at any rate relied on. Although polite references to them commonly implied that they would fulfil this purpose, in practice their effect would rather be to suggest to their users new reactions which they might test by experiment. If so, they would be a valuable stimulus to further discovery. Modern philosophers of science sometimes argue that one of the chief functions of a scientific theory is precisely to suggest experiments by which the limits of its applicability might be narrowed. It would be anachronistic, however, to attribute such subtleties to eighteenth-century chemists.

Since affinity tables were a thoroughly misleading guide in predicting the results of reactions, and since their unreliability was continually being pointed out during their history, it would be very surprising if chemists had continued to be interested in them for so long for no other purpose. Several authors make another purpose clear. It was inductive. The tables might reveal some general pattern which would make chemistry respectable and might even be expressed mathematically. The discovery of general, mathematical laws for chemistry that would correspond with Newton's laws of motion in mechanics was a remote but ever-beguiling dream for most eighteenth-century chemists. For example, the historian of the Academy (Fontenelle), in commenting on Geoffroy's paper, wrote:

That a body which is united to another, for instance a Solvent which has penetrated a Metal, should leave it to go and unite with another which is presented to it, is a thing of which the possibility would not have been guessed by the most subtle philosophers, and of which the explanation is still not too easy for them today. One imagines first that the second Metal fits the Solvent better than the former which has been abandoned by it, but what principle of action can one conceive in a better fit? It is here that the sympathies and attractions would come very much to the point, if they meant anything. But in the end, leaving as unknown that which is unknown, and keeping to certain facts, all the experiments of Chemistry prove that a particular Substance has more disposition to unite with one Substance than another, and that this disposition has different degrees ... This Table becomes in some sort prophetic, for if substances are mixed together, it can foretell the effect and result of the mixture, because one will see from their different relations what ought to be, so to speak, the issue of the combat ... If Physics could not reach the certainty of Mathematics, at least it cannot do better than imitate its order. A Chemical Table is by itself a spectacle agreeable to the Spirit, as would be a Table of Numbers ordered according to certain relations or certain properties.[128]

Similarly Douglas wrote in introducing his translation of Geoffroy's manuscript lectures:

These Affinities gave offence to some particular People, who were apprehensive they might be only Attractions disguised, and so much more dangerous, as some Persons of eminent learning had already cloathed them in seducing Forms; but they soon grew sensible that this was an unnecessary Scruple, and that M. Geoffroy's Table might well be admitted; which, if rightly understood, and carried on to the utmost Degree of exactness, might become a fundamental Law for Chemical Operations, and guide the Operator with success.[129]

A little later, Clausier wrote:

The affinities as we give them here are only an assemblage of experiments which having been repeated many times on the same materials, are as good as axioms in Chemistry ...[130]

In his *Éloge* of Bergman some thirty-five years later, Condorcet wrote:

M. Geoffroy, of this Academy, seems to be the first who thought of reducing to some general rules the phenomena observed constantly in chemical operations … M. Geoffroy thought of giving a table which contained, for the most important or best-known substances, the order of the force of affinity, according to which the others adhere to it. It was in accordance with the observations that this table had been formed; it was like a summary of the results which one could extract from it; it reduced to a small number of simple and general facts the mass, already very great, of chemical facts; it could serve to provide the explanation of several new phenomena, that is to say, to show their agreement with the facts then known. This table had the success which such an ingenious idea deserved, and since that time there is hardly any famous chemist who has not in his teachings or in his works given M. Geoffroy's table of affinities, corrected, augmented, brought to perfection.[131]

It was, of course, characteristic of the Enlightenment to want to reduce the apparent chaos of nature to comprehensible order by the application of human reason. The wish to reduce chemical phenomena to a tidy table, expressed by Condorcet and by Fontenelle before him, is a good example of this attitude of Enlightened philosophers.

Condorcet goes on to quote Bergman's estimate of the number of experiments required to complete his table. This passage stresses the aspect of the tables from which they appear as mere summaries of known facts, though the hope for 'general rules' (*règles générales*) is also mentioned.

A little later still, Lavoisier placed the emphasis rather differently:

The part of Chemistry most susceptible, perhaps, of becoming one day an exact science, is that which deals with affinities or elective attractions. M. Geoffroy, M. Gellert, M. Bergman, M. Scheele, M. de Morveau, M. Kirwan and many others have already assembled a multitude of particular facts, which now await only the place which has to be assigned to them; but the principal data are lacking, or at least those which we have are not yet either precise enough or certain enough to become the fundamental basis on which we must rest so important a part of Chemistry. Further, the science of affinities is to ordinary Chemistry what transcendental Geometry is to ordinary Geometry, and I have not thought it necessary to complicate by such great difficulties simple and easy Elements which will, I hope, be within the compass of a very great number of Readers.[132]

Professor Partington points out that many of the arguments used by the Irish chemist Richard Kirwan (1733–1812) in his *Essay on phlogiston* 'depend on apparent contradictions between reactions postulated by the anti-phlogistic theory and the regular order of substances'. He adds that Lavoisier (in the passage just quoted) 'expressly declined to enter upon a discussion of affinities, although he gave numerous tables of affinities in his book'.[133] However, Lavoisier and his supporters answered Kirwan in their notes to Mme Lavoisier's translation of Kirwan's *Essay*, and it is probably true that the subject was avoided in the *Traité* simply because it was rather too advanced for an elementary textbook and was in any case going to be thoroughly dealt with by Guyton. Lavoisier's desire to have chemistry placed on the foundation of a simple, general pattern is notable, and clearly part of his reason for writing what he claims to be an elementary textbook is to bring out that simple pattern.

In accordance with the purposes of affinity tables just described, attempts were made to determine quantitatively the magnitude of the force of affinity or chemical attraction. However, these attempts (which will be described in the next chapter) did not produce generally acceptable results. The hope that affinity tables would make possible the induction of general, preferably mathematical laws was not fulfilled either; for although, as we shall see, efforts were made to produce general laws of affinity, the results were not at all convincing. Yet it was these hopes which kept interest in affinity tables alive in spite of all disappointments. They were often criticized because they were not based on any theory; but that was precisely the point of them. They enabled the chemist to arrange the substances in an orderly pattern which did not imply any speculative theory, but depended on a sequence of experimentally observed properties. If any theory was to be conceived, it would be through the experimental results. The tables therefore seemed to offer an escape from the dilemma between the need to avoid unfounded speculation and the need to bring chemistry to the orderliness of a respectable and enlightened branch of philosophy.

Affinity tables and the classification of known substances

Affinity tables also performed an additional function. We are accustomed nowadays to thinking of the Periodic Table as the framework of chemistry, providing a list of all the elements in an order which was originally based on a gradation of observed properties, but was later seen to be based on a more fundamental characteristic, the number and arrangement of orbital electrons. Chemists in the eighteenth century had no such framework, but urgently felt the need of one. The reasons have already been explained: there was a practical need to classify the increasing number of substances so that the chemist could grasp some memorable and comprehensible pattern, rather than being faced with an apparently random jumble of phenomena, and there was an intellectual need to perceive a logical order. Affinity tables came near to supplying these needs. Like the Periodic Table, they began by simply showing an observed gradation of properties; but it was hoped that in time some more fundamental principle would be revealed.

Furthermore, the tables implied a distinction between simple substances, which were shown in the tables in their own right, and their compounds, which were not simple, but the result of the operation of the affinities or chemical attractions between simple substances that the tables recorded. Yet these simple substances, or commonly used ingredients found in the laboratory, were entirely different from the traditional lists of elements or principles, which might be approved by philosophers but were no longer relevant to the needs of chemistry. Nor had they anything to do with the mechanical explanations that were valuable in other branches of philosophy, academically respectable, and dutifully or ostensibly accepted by chemists, although contributing only indirectly to solving their real problems. Among mid-eighteenth century chemists, few remained who, like Macquer, offered a specific defence of the four Aristotelian elements as the ultimate components of matter. Even he accepted that other substances such as acids and alkalis were at least relatively simple; and even he hardly mentioned the Aristotelian elements when he was discussing particular chemical questions.[134] The headings of affinity tables, then, like the chapter headings of the business parts of textbooks of chemistry, constituted an unofficial list of simple substances that gradually became established during the eighteenth century, in contrast to those cited in the official doctrines of traditional philosophers that chemists still quoted in their introductory chapters.

In the early eighteenth century the word 'principles' still retained much of its early meaning of an insubstantial essence which gave a particular piece of matter its properties. The word 'elements' hovered between this meaning, ultimately derived no doubt from the Aristotelian concept of 'form', and the sense of the ultimate simple substances into which all chemical substances could be analysed. Some natural philosophers, like Boyle, believed that the most simple particles of matter had never been produced as separate samples in practice, and that all observed substances had been made of more complex particles. Yet to the practical chemist trying to understand the chemical changes which he observed in his laboratory the only helpful concept was that of simple substances from which compounds were formed and into which all compounds could be analysed. Whether they were simple substances in the absolute sense that it was not possible as a matter of ultimate philosophical principle to split them up into anything simpler—that is, in the sense which Boyle intended, and in which modern physicists are still hunting for quarks—was of no practical importance to chemists, though it might interest them out of curiosity.

As Siegfried and Dobbs have expressed it, 'the *a priori* schemes of "elements" or "principles" in use before the time of Lavoisier—such as the Greek tradition of Earth, Air, Fire and Water or Paracelsus's *tria prima* of Salt, Sulphur and Mercury—had obvious dissimilarities but this they had in common: that the elements were conceived more as metaphysical entities than as specific substances to be handled in the laboratory'.[135] Also, it was often not understood whether the substances which emerged from a reaction were components of the substances which had entered it or *vice versa*. Nevertheless, the writer of a textbook naturally had to describe and relate together 'specific substances to be handled in the laboratory', whatever he thought of their composition and nature.

Friedrich Hoffmann (1660–1742), Stahl's colleague at Halle, argued in his textbook that neither the Aristotelian elements, nor Salt, Sulphur and Mercury, nor acid and alkali, nor rare and dense, nor fixed and volatile, could properly be regarded as principles, because they did not fulfil the required conditions. Those were: (1) that they should be completely simple '*prima*' and not composite; (2) that they should be clear, distinct, and capable of being conceived and understood, not imagined; and (3) that all phenomena, as far as reality and the weakness of our understanding allow, can be deduced from them.[136] Thus he is well on the way to the kind of empirical definition of simple substances which suited chemistry, though he still insists that they should be completely simple in the philosophical sense, and he does not actually say that principles have to be visible and palpable substances that one could produce in a bottle, so to speak. In fact he went on to discuss mechanical explanations, which, as was academically respectable at the time, he said he preferred; but they did not figure very much in the main part of the book, where he was describing actual chemical phenomena.

The standard definition of elements or principles used by chemists during the eighteenth century does not vary very much. Sometimes in fact no distinction is made between them. Peter Shaw (1694–1763), for example, in the Glossary printed at the beginning of his *Chemical lectures* (delivered in 1731, 1732, and 1733) under the heading 'Elements' just put 'See Principles'. Under 'Principles' he put simply 'Are the different simple Matters whereof a Body is composed.'[137] Most writers gave a definition which meant the same, although it is usually expressed in many more words.

Spielmann, in his *Institutiones chemiae*, is one of those who did make a distinction between principles, as the constituent parts, immediate or remote, of substances and the product of their decomposition, and elements, as the 'natural bodies, the principles of which are homogeneous'. In other words, he is unusual in defining principles in a way fairly close to Lavoisier's later empirical definition of simple substances, but uses the traditional philosophical definition for elements. He also adopts Becher's mechanical explanation in terms of a hierarchy of compound particles in order of simplicity, implying that the elements are substances consisting of particles which are unmixed and entirely of the same kind, and that principles consist of groups of particles of the same kind.[138] Such speculative mechanical explanations were, however, generally out of fashion among chemists by his time, at least in Paris, which was at the forefront of chemistry.

A more common approach was still followed as late as 1796 by William Nicholson in his *First principles of chemistry*, in which he adhered to the obsolete phlogiston theory of combustion and managed to reject Lavoisier's theory. He wrote:

> Principles which cannot be subdivided by art, are called elements, or first principles, and the principles made up of these, are called secondary principles. Some writers carry this order still farther; but it must be confessed that no means have yet been devised to shew unequivocally whether any such subordination of principles exists. We may indeed discover the component parts of bodies; but we know nothing of their arrangement.[139]

Nevertheless, what chemists needed to make sense of their observations was a mental picture of substances that could be produced in visible form, and had constant weight, combining together to form compounds or separating to form new compounds, the compounds having constant composition by weight. The picture of matter as consisting of various kinds of invisibly small particles which each had constant weight and properties was indeed helpful; but chemists' needs were derived from more vulgar origins than the Aristotelian philosophy or the mechanical philosophy. For example in mining practice, when an ore was assayed it was necessary to assume that one sample of ore had the same composition by weight as any other sample, so that the proportion of valuable metal extracted from the sample would be the same as the proportion of such metal in the total quantity of ore available for mining. Otherwise assaying would be pointless. That also implied conservation of the weight of each of the simpler substances of which the ore was composed (which is not the same as conservation merely of the total weight), and indeed a conception of the ore as being composed of the substances extracted from it by analysis. They would bear little relation to the traditional doctrine of elements or principles. Furthermore, precise weighing was a normal practice in assaying. The assumptions and the practice were justified by experience.

Similarly in the preparation of medicines, an apothecary following a recipe would implicitly assume that the weight of each ingredient would be constant and that the composition of any sample of the final preparation would be the same as that of any other sample. After all, it was vitally important that the composition of the prepared medicine should be precisely right, so that it would cure the patient and not poison him. The apothecary's scales were an essential part of his equipment, and regularly used. Both the assayer of ores and the apothecary would naturally form a conception of a pure substance and of intruding impurities which had to be removed as far as possible, even if they did not attain purity in practice. Thus if we read between the lines it is clear that practising chemists in the eighteenth century implicitly assumed conservation of mass and constancy of composition, even though nobody stated them explicitly before Lavoisier.

What happened, then, was that chemists evolved their own operational list of simple substances, which served their purpose much better than the traditional notions of elements and principles could have done. It is a case of a new set of categories—in essence an important part of a new paradigm or exemplar—growing up as an independent alternative to the traditional set, which gradually faded away as the new generation ceased to use it, rather than of a revolutionary overthrow of the old set.

Hence one commonly finds in textbooks of the early and middle eighteenth century, after an introductory section on such topics as the aims of chemistry, the nature of principles, and the chief chemical processes, that the author makes his own selection from the Aristotelian elements and the *tria prima*, sometimes including in addition Acid and Alkali or the like. For this there were seventeenth-century precedents. For instance J. J. Becher (1635–*c*.1682) considered that all substances were composed of water and three earthy principles, which he developed from the *tria prima*.[140] Becher also counted air as a principle. Thus he managed to fit his scheme into both the Aristotelian and the Paracelsan frame. Stahl's phlogiston was a development of Becher's, which was his second earthy principle, and might therefore be classified among the earths. Lemery recognized three active principles (spirit or mercury, oil or sulphur, and salt), and two passive principles water and earth).[141]

Among eighteenth-century authors, Geoffroy counted as principles Fire, Water, Earth, Salt, and Oil or Sulphur.[142] Alberti dealt with Water, Air, Fire, and Earth as elements.[143] Rothen classified 'chemical productions' as saline, sulphureous, and earthy.[144] Spielmann says that the elements are water, the principle of fluidity; the earth, which is the principle of solidity and dryness (rightly called by Becher 'vitrescible'); and the inflammable principle phlogiston, which is also an earth, because it is dry, and is sometimes called 'sulphur' or 'fire'.[145] Cartheuser wrote that the principles, or 'subtle, heterogeneous and sensible parts into which substances are generally resolved by chemical analysis' were commonly called Salt, Sulphur, Water, and Earth.[146] Even later, Baumé's *Manuel de chymie* has sections on Fire, Phlogiston, Water, and Earth.[147]

These schemes were produced by chemists when they were adopting the posture of philosophers, to set appropriately respectable academic theories at the mastheads of their textbooks. They are based on the traditional belief that compounds showed properties which are a blend of the properties of the elements or principles composing them, so that those elements or principles could be identified from the properties of the compounds. Naturally, the schemes often conflicted with the practical requirement of listing the real substances which were actually found in the laboratory to form compounds or to be obtained by their analysis, and which generally had quite different properties from their compounds. Superficially the new, operational list was derived by making the names of traditional principles plural, that is by making them into classes of specific substances. Earth becomes Earths; Salt, Acid, and Alkali become plural (and acids and alkalis are no longer included among salts); Water holds its place in the list until it is shown to be a compound; and Air becomes plural as different gases or elastic fluids are discovered.

Hence it is not surprising to find that the columns of affinity tables also fall naturally into groups. The grouping owes much to the traditional division of substances, as modified to provide a classification which could take in specific substances to be handled in the laboratory; but it is adapted so that substances are grouped according to what they combine with, and so that neighbouring columns have similar contents and can be compared. The compiler of a table did not need to discuss whether the substances at the heads of his columns were elements or not, or to argue whether they included all the elements there were. Clearly in most of the tables there are substances or classes of substances which the compiler did not claim to be elementary. Apart from vague titles of organic substances such as Clausier's 'gums, jellies, balsams, mucilages, soap', several supporters of the phlogiston theory included metals, since they were regarded as distinct substances in normal use, even though they were supposed to be considered compounds of their calces with phlogiston. Again, numerous tables include both sulphur and vitriolic (sulphuric) acid, even though their compilers must have believed that one was a compound of the other. Phlogistonists believed that sulphur was a compound of the acid with phlogiston, while others believed that the acid was a compound of sulphur and something else.

Further, many compilers pointed out themselves that their tables were incomplete. It was the freedom from having to settle theoretical questions of this kind—whether their lists of substances were exhaustive, whether all the substances on them were elementary or not, which was a compound of which—that enabled them to produce tables which simply brought into order and pattern the usages of practical chemistry.

Geoffroy, for instance, has first Acid Spirits and the three mineral acids; then Absorbent Earth, Fixed and Volatile Alkalis, and Metallic Substances; then Sulphur, along with six individual metals; and finally water. At first sight Sulphur seems out of place among the metals; but the basis of the grouping is clear from the similarity of the columns within each of the groups. Sulphur goes with the metals because, like them, it is shown as combining with metals. In other words, substances are grouped on the basis of similar observed chemical properties. Acids are shown as having *rapport* with metals, and (except for Marine (hydrochloric) Acid) with Alkalis; and Absorbent Earth, Alkalis, and Metallic Substances as having *rapport* with acids (and in the case of Fixed Alkali (potassium or sodium hydroxide) with Sulphur). Water is in a class of its own, which might have been labelled 'miscellaneous', having *rapport* with Spirit of Wine (alcohol) and Salt. Grosse added *Aqua Regia* (a mixture of nitric and hydrochloric acids) to the acids, and Tartar (potassium hydrogen tartrate) in the 'miscellaneous' section.

A different approach is represented by Clausier, who after Fire, Water, Air, and Salts in general has a section for acids (including vinegar), alkalis, and neutral salts; then one for the sulphur principle (i.e., approximately, phlogiston), which also includes common sulphur, bitumen, oil, spirit of wine and white of egg; and then one for metals. Gellert, with his emphasis on metallurgy, puts first the five different earths with which he replaces Geoffroy's single 'Absorbent Earth'. His table seems to divide naturally into a section for earths, followed by alkalis and acids; a section comprising Nitre (potassium nitrate), Liver of Sulphur (potassium polysulphides), and Metals; and finally Glass on its own. The reason for this arrangement is evident from the resemblances between neighbouring columns within each section. The columns for earths all show, as combining with earths, calces (oxides or sometimes other stable solid salts) of metals or other earths near the top, and only in Column 5 (for Calcareous (i.e. chalky) Earth) do the acids come in, in the lower part. The next twenty-two columns consist mostly of metals; but the alkalis like Calcareous Earth have the acids in the lower part of their columns, and in addition have the Inflammable Principle (that is, more or less, phlogiston) at the bottom. The acids and nitre have in their columns metals followed by the Inflammable Principle. Sulphur, Liver of Sulphur, and the twelve metals have nothing but other metals. Finally, glass has nothing in its column but the calces of metals.

Most French tables were closer to Geoffroy's arrangement. Limbourg has first acids (including 'Vegetable Acid', which is the more enlightened name for refined vinegar or acetic acid), and then alkalis, Absorbent Earth and Aluminous Earth, as substances which combine with acids as well as with each other. Then come Lime (calcium oxide), Phlogiston, Sulphur, Liver of Sulphur, Slaked Lime (calcium hydroxide), Arsenic, Orpiment (arsenic trisulphide), Spirit of Wine (ethyl alcohol), Ether, Soap, Amber, and Water, forming a 'miscellaneous' section of substances which combine with a variety of other substances, and then eleven metals. Rouelle added to Geoffroy's table merely Spirit of Wine, on the right next to Water, and a division of metallic substances into Solar and Lunar. Demachy's table might be divided into Acids (including Vinegar and *Aqua Regia*), then Earth, Alkali and Borax (sodium tetraborate), then Metallic Substances, then a 'miscellaneous' section consisting of water,

Spirit of Wine and Fire, and finally Arsenic, Sulphur, Liver of Sulphur, and Metals. Sage's table is not unlike Geoffroy's but has Phosphoric Acid and Vinegar in the section for acids, Liver of Sulphur between Absorbent Earth and Metallic Substances, no separate metals except Mercury, and Phlogiston last but one before Water.

Bergman's much enlarged table falls quite clearly into five sections: (1) Acids; (2) Alkalis and Earths; (3) Water and Air; (4) a miscellaneous section, made up of Phlogiston, Sulphur, Liver of Sulphur, Spirit of Wine, and Oils; and (5) Metals. It includes for the first time a wide range of both organic and mineral acids, though without any indication of their composition; it includes Aerial Acid (Bergman's name for what would now be called carbon dioxide) among the acids; and it includes in what I have here called the 'miscellaneous section' substances which were thought to contain much phlogiston. This division is made explicit by Berkenhout in his simplified version of Bergman's table, in which he has four sections headed 'Acids'; 'Alkali and Earths'; 'Inflammibles &c'; and 'Metallic Substances'. He includes Water and Air under 'Inflammibles &c'. The substances included in this 'miscellaneous section', being those substances which do not fit easily into any of the general categories and have peculiar properties of their own, naturally tend also to be those which are of fundamental importance.

Nicholson's version of Bergman's table, in much the same way as Berkenhout's, has the general headings 'Acids', 'Alkalis and Earths', 'Combustible Substances and Water' (with rather more accuracy than Berkenhout), 'Phlogiston and Metals', and 'Metallic Substances'. The pages headed 'Metallic Substances' are merely a continuation of the one headed 'Phlogiston and Metals', and not a separate section. The reason for the omission of the column for air is given in a note: 'in the anti-phlogistic theory, the column intitled Phlogiston being taken in the reverse order, will express the elective attraction of Vital Air' (that is, of what was called by others oxygen). This is a neat illustration of the way in which affinity tables can represent the observed facts without commitment to any particular theory of composition.

Gergens and Hochheimer show three separate tables, the first divided into sections headed 'Mineral Acids', 'Vegetable Acids', 'Animal Acids', and (in a single column) 'Acid of Air' (i.e. what would now be called carbonic acid.) The second is divided into sections headed 'Alkalis', 'Earths', and 'Noble Metals', and the third into two sections headed 'Base Metals' and 'Various Substances'. This last, miscellaneous section comprises Water, Atmospheric Air, Phlogiston, Fire, Sulphur, Liver of Sulphur, Spirit of Wine, Ether, Essential Oils, and Fatty Oils.

To sum up, then, it appears that although the authors of affinity tables started from the old classification of substances according to the principles which were supposed to be their essence, they adapted these classifications so that substances were arranged according to what they combined with. Water and air sometimes appear by themselves, but tend to come in a separate section of important substances that do not fit into other categories. Salts appear as a general category in one or two early examples; but acids and alkalis are necessarily separate, since they combine with different kinds of substances, and the fact that they are salts in the old sense of the word is soon forgotten. Since neutral salts are a product of the combination of acids and alkalis, they are not shown separately in most tables of substances which combine, except to show that they dissolve in water.

The original single entry for 'Absorbent Earth' is expanded in the German tradition to several columns for different kinds of earth; and since they combine with much the same substances as alkalis they come to be grouped with alkalis. Metals are shown separately from the first, though there is one table (Sage's) in which the only metal shown at the heads of its own column is mercury, which in the old theory was the metallic principle. Sulphur is sometimes included among the metals, because like them it combines with various metals, and sometimes in a group which suggests that it represents the principle of inflammability. In Bergman's table, which was by far the most influential in the last quarter of the eighteenth century, inflammable substances appear in a group by themselves. The calces of metals are sometimes shown rather than the metals themselves in the body of the tables, in accordance with the theory that metals were not simple substances but compounds of phlogiston and their respective calces; but Bergman was the first, in the second version of his table, to take the logical step of placing the calces instead of the metals at the head of their columns.

Rouelle, Lavoisier's teacher, in the paper of 1744 which brought about the adoption of the phrase 'neutral salts', included a table classifying neutral salts in six sections according to their crystalline shapes, but also showing the acids and bases from which they were formed.[148] It is noticeable that the acids and bases are represented by their symbols in this table. Although the symbols were used in manuscripts as a form of shorthand they were seldom seen in print at that time except in Geoffroy's table, of which, as we have found, Rouelle himself issued a modified version.

A natural development from the columns for acids and alkalis in affinity tables was a table divided into squares with the acids along the top, the alkalis down the sides (or vice versa), and the neutral salt produced by the combination of each acid and alkali in the square opposite them. The earths and metals might be added to the alkalis. Such tables were produced by Guyton de Morveau, Berkenhout, and Trommsdorff, for instance.[149] Bergman's diagrams of double decompositions showed the products in symbols, and tables developed from his, such as the tables of

Weigel, also sometimes showed the names of the products.

It was Bergman who devised a binomial system of nomenclature for salts, giving each a name formed from the acid and the base composing it. He was influenced by his teacher Linnaeus, who had devised a binomial system of names for species of living creatures, based on his system of classification. Bergman in his turn influenced Guyton de Morveau in the work which led to the new nomenclature devised by the latter in collaboration with Lavoisier, Berthollet, and Fourcroy.[150]

Lavoisier and the autonomy of chemistry

Chemists thus found themselves in a dilemma. If they kept to the definition of an element or principle which satisfied traditional natural philosophy, that it was a substance that was simple in the ultimate sense that it could never conceivably be divisible into anything simpler, or indeed that it consisted of particles that could never be split up into simpler particles, then the substances that they had to treat as simple in practice obviously did not qualify as elements. On the other hand, as chemistry had attained autonomy and evolved its own distinctive concepts to fulfil its own requirements rather than those of physics, it had as we have seen developed its own unofficial list of substances that it had been found empirically useful to treat as simple substances. Lavoisier cut the Gordian knot. In defining an element as 'the last point that analysis is capable of reaching', Lavoisier deliberately avoids the question whether the chemical elements were ultimately simple. For the progress of practical chemistry it was better avoided. His new definition legitimizes chemists' own list of simple substances, for either they now fall within the new definition and qualify as elements, or else it is within the power of chemists to analyse them and show the elements of which they are composed.

Lavoisier's definition of 'simple substances' exerted great influence not because it was entirely original, but because it was made a central feature of a new system, depending on his oxygen theory, which set chemistry on an orderly, comprehensible basis, and because it was accompanied by an actual list of what he considered were simple substances. This made his definition more effective and his meaning clear, even if the list was open to criticism. Indeed, part of its value in the long run was that it could be criticized in detail, and so revised and improved. It was also more effective because it was expressed in the new nomenclature. Consequently, the names of the substances indicated their composition in terms of Lavoisier's broad scheme of chemistry. In view of his criticisms of affinity tables, it comes as a surprise to find that his list of simple substances is essentially a development of them.[151] However, as the headings of the columns of affinity tables necessarily embodied the chemists' empirical, unofficial list of simple substances that Lavoisier now made official, that should not be a surprise.

In fact, the second part of the first volume of his *Traitée elémentaire de chimie* is virtually an extended affinity table with a commentary. It is, however, an affinity table with a new purpose, for it is a means of conveying Lavoisier's new scheme of chemistry. Apart from a few fundamental substances, such as Heat and Oxygen, on which the whole structure is founded, all substances in inorganic chemistry fitted into the structure, within which non-metals combine with oxygen to form acids, metals combine with oxygen to form bases, and bases combine with acids to form salts. Lavoisier made it clear that his tables form a summary of the facts of chemistry, arranged according to this pattern. The names of the products of the combination of each substance with that at the head of the column are also shown.

At the beginning of this section of the book is a *Tableau des substances simples*, the famous so-called 'list of elements'. Naturally, it reads very like a list of the headings of the columns of an affinity table drawn up according to Lavoisier's system. As we have seen, it follows from their purpose that the headings of the columns of affinity tables constitute a list of substances that chemists found themselves treating as simple substances in laboratory practice.[152] Lavoisier's list is in four sections, the first labelled 'Substances which belong to the three kingdoms and which one can regard as the elements of substances'. These are Light, Heat (Caloric), Oxygen, Nitrogen, and Hydrogen. The section resembles the 'miscellaneous' sections of earlier tables, which had often included the most important basic substances; but in Lavoisier's system the section has a new significance as the basis of the whole scheme. Water has been replaced by Oxygen and Hydrogen.

The second section contains the 'Simple, non-metallic, oxidizable and acidifiable substances'. This is equivalent both to the section for acids and to the section for inflammable substances in Bergman's table, since in Lavoisier's system the acids (which would nowadays be called acidic oxides) were the product of the oxidization of non-metallic inflammable substances. The third section is for metals, which on oxidization produce bases, and the fourth section is for what had previously been called 'earths'. Lavoisier does not include alkalis, because he does not believe them to be 'simple substances'.

After this table of simple substances one might expect Lavoisier to give a list of possible combinations in order of affinity for each of the simple substances. In fact, however, he gives lists of combinations (not all in order of affinity) only for Oxygen, Nitrogen, Hydrogen, Sulphur, Phosphorus, Carbon, and the various compound acid radicals. These last mostly correspond with Bergman's vegetable and animal acids. Of the simple substances, Light and Caloric are omitted, presumably because they do not enter into chemical compounds in the ordinary sense, and the metals and earths are omitted, presumably because their combinations with acids are completely covered by the tables for the acids. The combinations of metals with each other are not included, Lavoisier says, because the tables would be very voluminous and would depend on research which was not yet complete.[153]

It is evident that Lavoisier in this part of the book has worked from the tradition exemplified by affinity tables, and that he has not made a complete break with the past. Although his new scheme of chemistry was revolutionary in its effects, he has been able to accommodate it very largely within the conventional framework developed by chemists during the eighteenth century. Nevertheless, he has added Light, Heat, Oxygen, Nitrogen, and Hydrogen to the substances which usually appeared in affinity tables, and transformed what I have called the 'miscellaneous' section of the existing tables into a new role; he has extended the list of acids and made it clear that many acid radicals are compound; and above all he has fundamentally altered the significance of the conventional pattern by founding his tables on his new system of chemistry, which, indeed, he is using them to present. By producing an orderly system which emerges naturally from the observed data of chemistry he has succeeded in avoiding the dilemma between the inhibition against theoretical speculation and the need to find a reasonable pattern. Chemistry has come of age and gained its independence.

By the beginning of the nineteenth century the day of the affinity table was over, partly because the idea was incapable of further development and had not been a success, and partly because interest had shifted to other lines of investigation. Probably some people continued to use them simply as a convenient guide to the basic facts, and a few people spoke of them in a kindly way. The situation was summed up by Thomas Graham, writing in his textbook of 1842:

> Tables of this kind when accurately constructed may convey much valuable information of a practical kind, but it is never to be forgotten that they are strictly tables of the order of decomposition and of the comparative force or order of affinity in one set of circumstances only.[154]

Graham adds that circumstances may affect the order of decomposition. In speaking of affinity tables in the present tense he was perhaps already a little old-fashioned. Even today, with all the knowledge of atomic and molecular structures, thermodynamics, and wave mechanics which has been acquired in the past hundred and fifty years, no chemist would claim to be able to predict with certainty the result of an untried reaction. However, there is no denying that a table which would enable him to do so would be very handy. It had been an unfulfilled dream, but a beguiling one.

Notes

1. Geber, *Works*, trans. R. Russell, ed. E. J. Holmyard, London, 1928, pp. 181, 192, 194; Paracelsus, *Werke*, ed. Sudhoff, 14 vols, Munich and Berlin, 1922–33, Vol. XI, p. 365; Newton, *Opticks*, Dover reprint, New York, 1952, pp. 380–1; Glauber, *Works*, trans. Packe, London, 1688, Part I, p. 74; Stahl, Philosophical Principles of Universal Chemistry, trans. Shaw, London, 1730, p. 39.
2. E.-F. Geoffroy, 'Table des différens rapports observés en chimie entre différentes substances', *Mémoires de l'Académie Royale des Sciences*, 1718, pp. 202–12 (published 1719). See W. A. Smeaton, 'E. F. Geoffroy was not a Newtonian Chemist', *Ambix*, **18**, 1871, pp. 212–14.
3. Reproduced at the end of Demachy, *Recueil de dissertations*, Amsterdam, 1774.
4. J. Quincy, *Pharmacopée universelle raisonnée*, trans. Clausier, Paris, 1749, pp. 6–8.
5. C. E. Gellert, *Anfangsgründe zur metallurgischen Chymie*, 2nd edn, Leipzig, 1776 (1751), opp. p. 172.
6. William Lewis, *The new dispensatory*, London, 1753, p. x. See N. Sivin, 'William Lewis (1708–81) as a chemist', *Chymia*, **8**, 1962, pp. 63–88.
7. D. A. Rüdiger, *Systematische Anleitung zur reinen und überhaupt applicirten oder allgemeinen Chymie*, Leipzig, 1756, between pp. 266 and 267.
8. J. P. de Limbourg, *Dissertation sur les affinitès chymiques*, Liège, 1761, at end.
9. P. A. Marherr, *Dissertatio chemica de affinitate corporum*, Vienna, 1762, at end.
10. This table was printed without mention of the author in Diderot and d'Alembert, *Encyclopédie*, 35 vols, Paris, 1751–80, Vol. III of *Recueil des planches*, under heading '*Chimie, laboratoire et table des rapports*'. It was reprinted in Guyton de Morveau, *Elémens de chymie, théorique et pratique*, 3 vols, Dijon, 1777–8, Vol. I, opp. p. 90 (bound at end of Vol. III in British Library copy). It is there called '*Table des affinités*', and is said to be Geoffroy's table altered by Rouelle. In both publications the differences from Geoffroy's table, which are few, are marked with asterisks.
11. J. R. Spielmann, *Institutiones chemiae*, Strasburg, 1763; 2nd edn, Strasburg, 1776; French trans. by Cadet le Jeune, *Instituts de chymie*, Paris, 1770, Vol. I, p. 24.
12. John Warltire, *Analysis of a course of lectures in experimental philosophy*, 6th edn, London, 1769, appendix entitled 'Tables of the various combinations and specific attractions'.
13. J. R. Partington, *History of chemistry*, Vol. III, London, 1964, p. 99, refers to a publication of affinity tables by Demachy in *Précis d'une nouvelle table des combinaisons ou rapports*, Paris, n.d., but *c*.1770. I have not seen that, but no doubt it has the same table as that printed in Demachy, *Recueil de dissertations*, Amsterdam, 1774, at end. The table is there dated 1769.
14. Tessari, *Chymiae elementa in aphorismos digesta*, Venice, 1772, p. 10.
15. Sage, *Mémoires de chimie*, Paris, 1773, pp. 259–62.
16. Discussed in de Fourcy, 'Observations sur le tableau du produit des affinités chymiques', *Rozier's Observations sur la physique*, 1773, **2**, pp. 197–204. A note in this paper says that 'on trouve ce tableau chez Collard Graveur, quai de la Mégisserie, Image St Michel', but I have not been able to trace a copy. There is none in the Bibliothèque Nationale.
17. J. C. P. Erxleben, *Anfangsgründe der Chemie*, Göttingen, 1775, pp. 45–471.
18. This original version is in Torbern Bergman, 'Disquisitio de attractionibus electivis', *Nova Acta Regiae Societatis Scientiarum Upsaliensis*, 1775, **2**, 161–250;

trans. with an introduction by J. A. Schufle, *Dissertation on elective attractions*, New York and London, 1968.

19. This enlarged version is in Bergman, *Opuscula physica et chemica*, Uppsala, 1783, Vol. III.
20. John Elliot, *Elements of the branches of natural philosophy connected with medicine*, London, 1782, opp. p. 105.
21. Bergman, *A dissertation on elective attractions*, trans. anonymously by Thomas Beddoes, London and Edinburgh, 1785. Differences from Bergman summarized in my introduction to the 1970 repr., pp. xxxv-xxxvii.
22. J. Berkenhout, *First lines of the theory and practice of philosophical chemistry*, London, 1788, opp. pp. 138, 143, 146.
23. J. C. Wiegleb, *General system of chemistry*, trans. C. R. Hopson, London, 1789, at end.
24. W. Richardson, *The chemical principles of the metallic arts*, Birmingham, 1790, at end.
25. W. Nicholson, *The first principles of chemistry*, London, 1790, Tables VI-XI.
26. George Pearson, *A translation of the table of chemical nomenclature*, 2nd edn, London, 1799, Tables III and IV.
27. D. C. E. Weigel, *Grundriss der reinen und angewandten Chemie*, Greifswald, 1777, Tables III-VI at end of Vol. I.
28. J. C. Wiegleb, *Handbuch der allgemeinen Chemie*, 2 vols, Berlin and Stettin, 1781, Vol. I, pp. 322–41.
29. John Elliot, *Elements of the branches of natural philosophy connected with medicine*, London, 1782, opp. p. 105.
30. A.-L. de Lavoisier, *Traité élémentaire de chimie*, 2 vols, Paris, 1789, Vol. I.
31. P. Gergens and S. Hochheimer, *Tabellen über die chemische Verwandtschaft der Körper*, Frankfurt-am-Main, 1790.
32. F. A. C. Gren, *Systematisches Handbuch der gesammten Chemie*, 2 vols, Halle, 1787–90: table at end of Vol. II. Revised version of table in 2nd edn, 4 vols, Halle, 1794–6, Vol. IV (1796), pp. 143–216. Another version of the table in the abridged edition, *Grundriss der Chemie*, 2 vols, Halle, 1796–7, Vol. II, pp. 337–72.
33. J. B. Trommsdorff, *Darstellung der Saüren, Alkalien, Erden und Metalle*, Erfurt, 1800, Tables 9–12.
34. J. Black, 'Experiments upon magnesia alba, quicklime and some other alcaline substances', *Essays and observations, physical and literary*, Edinburgh, 1756, Vol. II, pp. 224–5.
35. Lesage, *Essai de chimie méchanique*, Geneva, 1762, pp. 50, 52.
36. Lewis, *Commercium philosophico-technicum*, London, 1763, pp. 577–90.
37. J. F. Meyer, *Chymische Versuche*, Hanover and Leipzig, 1764; French trans., 2 vols, Paris, 1765, Vol. II, p. 247.
38. A.-L. de Lavoisier, 'Mémoire sur l'affinité du principe oxygine avec les différents substances auxquelles il est susceptible de s'unir', *Mémoires de l'Académie Royale des Sciences*, 1782, p. 530.
39. MS of Cullen's lectures belonging to Dr W. A. Smeaton of University College, London.
40. MS of Cullen's lectures at the Wellcome Historical Medical Library.
41. J. Black, *Notes from Dr Black's Lectures on chemistry, 1767/8*, ed. D. McKie, published by ICI Ltd, 1966, pp. 161–5.
42. Macquer, *Elémens de chymie-théorique*, 2nd edn, Paris, 1753, pp. 161–5.

43. Macquer, ibid., p. 296.
44. E.-F. Geoffroy, *Mémoires de l'Académie Royale des Sciences*, 1718, p. 203.
45. E.-F. Geoffroy, 'Eclaircissement sur la table inserée dans les *Mémoires* de 1718', *Mémoires de l'Académie Royale des Sciences*, 1720, pp. 20–3; Demachy, *Recueil de dissertations*, Amsterdam, 1774, pp. 86–7.
46. C. F. Geoffroy, 'Observations sur la terre d'alun; manière de la convertir en vitriol, ce qui fait une exception à la Table des rapports en chymie', *Mémoires de l'Académie Royale des Sciences*, 1744, pp. 69–76.
47. Demachy, *Recueil de dissertations*, Amsterdam, 1774, pp. 94–9.
48. H. L. Duhamel du Monceau and Grosse, 'Sur les différentes manières de rendre le tartre soluble', II, *Mémoires de l'Académie Royale des Sciences*, 1733, pp. 260–72; and Grosse, 'Recherche sur le plomb', ibid, pp. 313–28, esp. pp. 320, 322.
49. Quincy, *Pharmacopée universelle raisonée*, trans. Clausier, Paris, 1749, pp. 1–58.
50. Demachy, *Recueil de dissertations*, Amsterdam, 1774, p. 130; Guyton de Morveau, *Encyclopédie méthodique*, Paris and Liège, 1786–9, p. 537.
51. Gellert, *Anfangsgründe zur metallurgischen chymie*, 2nd edn, Leipzig, 1776 (1751).
52. Demachy, *Recueil de dissertations*, Amsterdam, 1774, pp. 113–14; J. H. Pott, *Chymische Untersuchungen*, Potsdam, 1746; 2nd edn, Berlin, 1757, p. 3.
53. Macquer, *Elémens de chymie-théorique*, 2nd edn, Paris, 1753; A. Baumé, *Manuel de chymie*, 2nd edn, Paris, 1765 (1763), p. 7.
54. Gellert, *Metallurgic chemistry*, trans. I. S. (probably John Seifert), p. 81.
55. Demachy, *Recueil de dissertations*, Amsterdam, 1774, pp. 112–23.
56. Gellert, *Metallurgic chemistry*, trans. I. S., p. 186.
57. Loc. cit.
58. N. Sivin, 'William Lewis (1708–81) as a chemist', *Chymia*, **8**, 1962, pp. 63–88.
59. Lewis, *Commercium philosophico-technicum*, London, 1763, pp. iv ff.
60. Caspar Neumann, *Chemical works*, trans. Lewis, London, 1755; e.g. p. 172 for vitriolic acid.
61. Lewis, *Commercium philosophico-technicum*, London, 1763, pp. 577–90.
62. *Notes from Dr Black's Lectures on chemistry, 1767/8*, ed. D. McKie, published by ICI, Ltd, 1966, pp. 161–5.
63. Rüdiger, *Systematische Anleitung zur reinen und überhaupt applicirten oder allgemeinen Chymie*, Leipzig, 1756, pp. 248 ff.
64. Demachy, *Recueil de dissertations*, Amsterdam, 1774, pp. 124 ff.
65. J. P. de Limbourg, *Dissertation sur les affinités chymiques*, Liège, 1761, at end.
66. P. A. Marherr, *Dissertatio chemica de affinitate corporum*, Vienna, 1762, Praefatio (pages not numbered).
67. Marherr, ibid., pp. 16–20.
68. Marherr, ibid., p. 75.
69. Marherr, ibid., Praefatio, (Neumann and Macquer), p. 46 (Margraf), p. 17 (Plummer).
70. Guyton de Morveau, *Elémens de chymie, théorique et pratique*, 3 vols, Dijon, 1777–8, Vol. I, opp. p. 90 (see Note 10).
71. Diderot and d'Alembert, *Encyclopédie*, 35 vols, Paris, 1751–80, Vol. III of *Recueil des planches* (see Note 10).
72. J. R. Spielmann, *Institutiones chemiae*, 2nd edn, Strasburg, 1766 (1763); French trans. by Cadet le Jeune, *Instituts de chymie*, Paris, 1770, Vol. I, p. 24.
73. John Warltire, *Analysis of a course of lectures in experimental philosophy*, 6th edn,

London, 1769: appendix 'Tables of the various combinations and specific attractions'.

74. Demachy, *Recueil de dissertations*, Amsterdam, 1774.
75. Demachy, ibid., pp. 183–92.
76. Tessari, *Chymiae elementa in aphorismos digesta*, Venice, 1772.
77. Sage, *Mémoires de chimie*, Paris, 1763.
78. Sage, ibid., p. 259.
79. De Fourcy, *Rozier's Observations sur la physique*, 1773, **2**, pp. 197–204.
80. De Fourcy, ibid., p. 201.
81. Erxleben, *Anfangsgründe der Chemie*, Göttingen, 1775, pp. 45–471.
82. A. Baumé, *Manuel de chymie*, 2nd edn, Paris, 1765 (1763), p. 7.
83. J. R. Partington, *History of chemistry*, London, 1964, Vol. II, p. 679, quoting G. E. Stahl, *Ausführliche Betrachtung ... von der Saltzen*, Halle, 1723, p. 425.
84. Torbern Bergman, 'Disquisitio de attractionibus electivis', *Nova Acta Reg. Soc. Sci. Upsal.*, 1775, **2**, 161–250; *Opuscula physica et chemica*, Uppsala, 1783, Vol. III.
85. Bergman, *Dissertation on elective attractions*, London, 1970 (1785), p. 7.
86. Ibid., p. 10.
87. Ibid., pp. 10–12.
88. Ibid., pp. 14–16.
89. Ibid., pp. 18–29.
90. Ibid., pp. 29–31.
91. Ibid., pp. 37–9.
92. Ibid., pp. 41–2.
93. Ibid., pp. 44–5.
94. Ibid., pp. 47–8.
95. G. F. Rouelle, 'Mémoire sur les sels neutres', *Mémoires de l'Académie Royale des Sciences*, 1744, pp. 353–64.
96. Bergman, *Dissertation on elective attractions, London, 1970* (1785), p. 51.
97. Ibid., pp. 54–6.
98. Ibid., pp. 56–7.
99. Ibid., p. 65. Pointed out by Charles Ross, 'Studies in the chemical work of Claude Louis Berthollet', unpublished London University M.Sc. dissertation, 1933, Ch. III.
100. Bergman, *Dissertation on elective attractions*, London, 1970 (1785), p. 70.
101. C. W. Scheele, *Collected papers*, trans. L. Dobbin, London, 1831, p. 275.
102. Bergman, *Dissertation on elective attractions*, London, 1970 (1785), pp. 230–5.
103. Ibid., pp. 263–4.
104. Ibid., pp. 244–8.
105. Bergman, *Dissertatio chemica de diversa phlogisti quantitate in metallis*, Uppsala, 1782; *Dissertation on elective attractions*, London, 1970 (1785), pp. 268–70.
106. Schofield, 'Joseph Priestley, the theory of oxidation, and the nature of matter', *Journal of the History of Ideas*, 25, 1964, pp. 287–9.
107. Bergman, *Dissertation on elective attractions*, London, 1970 (1785), pp. 82–6.
108. He explains that this is the reason in the case of acids: ibid., p. 86.
109. Bergman, ibid., pp. 16–17.
110. Béguin, *Elémens de chymie*, Paris, 1615, pp. 167–8. See p. 34 above.
111. See W. P. D. Wightman, 'William Cullen and the teaching of chemistry', *Annals of Science*, **11**, 1955, pp. 154–65 and **12**, 1956, pp. 192–205; M. P.

Crosland, 'The use of diagrams as chemical "Equations"', *Annals of Science*, **15**, 1959, pp. 75–90.

112. Crosland, ibid., pp. 79–80.
113. Black, *Lectures on the elements of chemistry*, 2 vols, Edinburgh, 1803, Vol. I, pp. 279–80; Crosland, ibid., pp. 80–1.
114. Crosland, ibid., p. 83.
115. Crosland, ibid., pp. 85–9; Black, *Notes from Dr Black's Lectures on chemistry*, 1966, pp. 164–5. (Reproduced in Appendix I, No. 32.)
116. Black, *loc. cit.*
117. D. C. E. Weigel, *Grundriss der reinen und angewandten Chemie*, Greifswald, 1977, Tables III-VI at end of Vol. I.
118. The term *'appropriatio'* for the drawing together by a third substance of two substances which would not otherwise combine seems to have been introduced by J. F. Henkel, *Mediorum chymicorum non ultimum, conjunctionis primum, appropriatio*, Dresden and Leipzig, 1727. Henkel does not, however, seem to have used the phrase *affinitas appropriata* himself, as is stated by Professor Partington, *History of chemistry*, Vol. III, London, 1962, p. 86.
119. J. C. Wiegleb, *Handbuch der allgemeinen Chemie*, 2 vols, Berlin and Stettin, 1781, Vol. I., p. 19.
120. John Elliot, *Elements of the branches of natural philosophy connected with medicine*, London, 1782.
121. John Elliot, 'Observations on the affinities of substances in spirit of wine', *Philosophical Transactions of the Royal Society*, **76**, 1786, p. 155.
122. Elliot, *Elements of the branches of natural philosophy connected with medicine*, London, 1782.
123. P. Gergens and S. Hochheimer, *Tabellen über die chemische Verwandtschaft der Körper*, Frankfurt-am-Main, 1790.
124. F. A. C. Gren, Systematisches Handbuch der gesammten Chemie, 2 vols, Halle, 1787–90: table at end of Vol. II. Revised version in 2nd edn, 4 vols, Halle, 1794–6, Vol. IV/1796), pp. 143–216. Another version in the abridged edn *Grundriss der chemie*, 2 vols, Halle, 1796–7, Vol. II, pp. 337–72.
125. J. B. Trommsdorff, *Darstellung der Saüren, Alkalien, Erden und Metalle*, Erfurt, 1800, Tables 9–12.
126. A.-L. de Lavoisier, 'Mémoire sur l'affinité du principe oxygine avec les différentes substances auxquelles il est susceptible de s'unir', *Mémoires de l'Académie Royale des Sciences*, 1782, p. 530.
127. Private communication from Dr J. Crellin.
128. *Histoire de l'Académie Royale des Sciences*, 1718, pp. 35–7.
129. E.-F. Geoffroy, *A treatise of the fossil, vegetable and animal substances that are made use of in physick*, trans. Douglas, London, 1736, p. xii.
130. J. Quincy, *Pharmacopée universelle raisonée*, trans. Clausier, Paris, 1749, p. 5.
131. Condorcet, *Oeuvres*, 12 vols, Paris, 1847–9, Vol. III, pp. 151–2.
132. A.-L. de Lavoisier, *Traité elémentaire de chimie*, Paris, 1789, Vol. I, pp. xiii-xiv.
133. J. R. Partington, *History of chemistry*, London, 1964, Vol. III, p. 664, referring to R. Kirwan, *Essay on phlogiston*, London, 1784 (French trans., *Essai sur le phlogistique*, by Mme Lavoisier, with notes by Guyton, Lavoisier, Laplace, Monge, Berthollet, and Fourcroy, Paris, 1788; 2nd English edn, with trans. by William Nicholson of the notes from the French edn, London, 1789).
134. P.-J. Macquer, *Elémens de chymie théorique*, Paris, 1749, pp. 12–15, and *Dictionnaire de chymie*, 2 vols, Paris, 1766, article *Elémens*.

135. Robert Siegfried and Betty Jo Dobbs, 'Composition, a neglected aspect of the chemical revolution', *Annals of Science*, **24**, 1968, p. 276. See also Marie Boas Hall, 'The history of the concept of element', in D. S. L. Cardwell (ed.), *John Dalton and the progress of science*, Manchester, 1968, pp. 21–39.
136. F. Hoffmann, *Chemia rationalis et experimentalis*, Leiden, 1748, pp. 2–7.
137. P. Shaw, *Chemical lectures*, London, n.d., pp. x and xiii.
138. J. R. Spielmann, *Institutiones chemiae*, 2nd edn, Strasburg, 1766 (1763); French trans. by Cadet le Jeune, *Instituts de chymie*, Paris, 1770, Vol. I, pp. 13–14.
139. W. Nicholson, *The first principles of chemistry*, London, 1790; 3rd edn, London, 1796, p. 75.
140. J. J. Becher, *Physica subterranea*, 4th edn, Leipzig, 1738 (1667), pp. 19, 53, 60, 70.
141. N. Lemery, *Cours de chimie*, Paris, 1756 (1675), p. 3.
142. E.-F. Geoffroy, *A treatise of the fossil, vegetable, and animal substances that are made use of in physick*, trans. G. Douglas, London, 1736, pp. 10–19.
143. M. Alberti, *Physices propositiones fundamentales*, Halle, 1721, pp. 38–53.
144. G. Rothen, *A synopsis, or, short analytical view of chemistry*, trans. Alexander Macbean, London, 1743 (1717), p. 4.
145. J. R. Spielmann, *Institutiones chemiae*, 2nd edn, Strasburg, 1766 (1763); French trans. by Cadet le Jeune, *Instituts de chymie*, Paris, 1770, p. 20.
146. J. F. Cartheuser, *Elementa chemiae medicae dogmatico-experimentalis*, Halle, 1736, p. 6.
147. Baumé, *Manuel de chymie*, 2nd edn, Paris, 1765 (1763).
148. Rouelle, *Mémoires de l'Académie Royale des Sciences*, 1744, p. 364, Plate XVI.
149. Guyton de Morveau, *Elémens de chymie*, 3 vols, Dijon, 1777, Vol. I, pp. 90 ff; *First lines of the theory and practice of philosophical chemistry*, London, 1788, opp. p. 154; J. B. Trommsdorff, *Darstellung der Saüren, Alkalien, Erden und Metalle*, Erfurt, 1800, Table IV.
150. Bergman, 'Meditationes de systemate fossilium naturali', *Nova Acta Reg. Soc. Sci. Upsaliensis*, **4**, 1784, pp. 63–128. See M. P. Crosland, *Historical Studies in the Language of Chemistry*, London, 1962, pp. 144–92.
151. A.-L. de Lavoisier, *Traité élémentaire de chimie*, Paris, 1789, Vol. I, pp. xvii-xviii. The resemblance between Lavoisier's way of arranging the bases which combine with each acid 'in the order of their affinities for this acid' in Part II of the *Traité élémentaire*, and Bergman's arrangement of the bases in his tables for the wet way was pointed out by J. M. Stillman, *The story of alchemy and early chemistry*, New York, 1960 (1924), p. 510.
152. A.-L. de Lavoisier, ibid., Vol. I, p. 192. For a criticism of my earlier publication on this topic, in which I did not emphasize sufficiently that the significance of the first section of Lavoisier's list of simple substances becomes very much wider than that of a merely 'miscellaneous' section (A. M. Duncan, 'The functions of affinity tables and Lavoisier's List of elements', *Ambix*, **17**, 1970, pp. 28–42), see C. E. Perrin, 'Lavoisier's Table of the elements: a reappraisal', *Ambix*, **20**, 1973, pp. 96–8.
153. Lavoisier, ibid., p. 230.
154. T. Graham, *Elements of chemistry*, London, 1842, p. 187.

5

Classification, quantification, and explanation

Types of explanation

If eighteenth-century chemists used such words as 'affinity' and 'attraction' not to indicate any theoretical explanation, but simply to mean the observed tendency of certain substances to combine, and if their affinity tables were intended chiefly to set out in an orderly way the observed facts from which they hoped some general principles might be derived, though without entertaining any theoretical presuppositions about the nature of affinity or elective attraction, we may well ask what the point was of their using the words at all. That involves asking what eighteenth-century experimental philosophers were trying to achieve, or what kind of result they would have counted as a success.

When a philosopher claimed to show that a complete system embracing all knowledge could be based on sure principles, as when Descartes erected his philosophy on the foundations of 'I think, therefore I am' and his proof of the existence of God, or when Locke established to his satisfaction that all the human mind contains must be derived or developed from sense impressions, the success of such achievements was probably measured by the feeling in their readers' minds that they had been given understanding of the nature of things. Natural philosophers, however, demanded not necessarily that their results should give this sense of grasping and understanding the whole basis of human knowledge, but that they should explain phenomena. Whereas the pure philosopher could take his data and his arguments entirely from his reason—and in most cases believed fervently at this period that the reason could solve all problems—the natural philosopher was obliged at least to save the phenomena, and so was obliged at least to find out what they were.

At the roots of any system of thought there must, of course, be unproved assumptions. Before any propositions can be proved, there must be axioms from which to prove them. The assumption that observation is better evidence than reasoning, where the two contradict each other, and the choice of what rules of argument are to be accepted, what kind of statement is sensible and what kind is nonsense, and what should be taken as obvious without need of proof, have to be made as matters of faith or personal preference before reasoned arguments can begin. For instance, as we have seen, Bergman assumed without any particular evidence that the order of elective affinities was essentially constant if other factors did not interfere. He had to assume either that it was so or that it was not so before he could begin to analyse the workings of affinity in practice. Similarly, assumptions have to be made about what constitutes a satisfactory explanation before the task of looking for one can usefully begin.

The tradition inherited from Greek and medieval philosophy was on the whole that phenomena were explained if they could be shown to be instances of a general principle at work. Some explanations of this kind were produced in the eighteenth century. For instance, when Gowin Knight reduced all phenomena to instances of the working of forces of attraction and repulsion, he clearly felt that this was a great achievement in itself. Nevertheless, there is no question of its leading to crucial experiments which might verify the theory. Nor is there any suggestion that it might lead to predictions of the results of new combinations of substances, or to the discovery of new reactions or compounds, let alone to unexpected results which might require the theory itself to be modified.

The practising chemist, however, was unlikely to be satisfied with such purely intellectual achievements, though he might respectfully admire them as a man of reason. He needed a theory that would help him understand the things that he saw and heard and smelt happening in his own apparatus, and, if possible, predict what was going to happen. In addition, as we have seen, he was extremely wary of theories that were not supported by experimental evidence.

Two main kinds of explanation were acceptable by the conventions of the age in the other physical sciences. One was the mathematical kind of explanation. For instance, Newton's law of gravitation explained why bodies fell towards the earth's surface in the sense that it enabled the philosopher to calculate the force acting on them towards the centre of the earth by a general mathematical law which applied equally to any pieces of matter anywhere in the universe. It was an explanation because it subsumed the particular phenomenon under a more general law, and gave quantitative and not merely qualitative explanations. Also, its stark, simple mathematical form, using algebraic symbols which were defined completely, or so it seemed, by precise directions for measuring them, and involving no occult or supernatural beliefs or unsupported speculations, appealed particularly to the taste of enlightened philosophers. Chemists, as I have mentioned, hoped that mathematics could be applied to chemistry in a similar way, and therefore tried in the latter part of the eighteenth century to measure the force of affinity or chemical attraction in numbers. Beyond that was the hope, though it remained distant, of finding mathematical laws of chemistry.

The other kind of explanation which was acceptable in the physical sciences was the mechanical explanation, which ideally of course would be expressible mathematically, so that the two kinds of explanation were related. In fact, they sometimes seemed to be opposed in particular cases, and different kinds of mechanical explanation often seemed to be opposed because they rested on different assumptions about what was acceptable. For example, explanations depending on the properties of some substance or substances, often otherwise unobservable, such as the matter of light or the matter of heat (which were commonly accepted in the eighteenth century, and are now known as imponderable fluids), appealed to those who found no difficulty in believing in such substances. On the other hand, explanations depending on the shapes and sizes of unobservable particles and their interaction (which had been more readily accepted in the seventeenth century) appealed to those who found unobservable substances incredible.[1] Cartesians thought action at a distance an impossibility. Yet such explanations are not necessarily irreconcilable, for substances like the matter of light and the matter of heat could be conceived as consisting of particles—as indeed they generally were in the eighteenth century—so that their properties would eventually depend on the behaviour of the particles, and even action at a distance might be explained by the mechanisms of an ether.

Mechanical explanations were satisfying because they enabled the natural philosopher to picture the invisible workings of nature as being like the workings of things that he could see and with which he was familiar. For instance, the notion that acids are acid because their particles have sharp spikes was of this kind. In the eighteenth century such a picture came to be considered as rather crude, and incapable of explaining chemical changes, partly because it showed only why acids acted on other substances when their particles were in contact, and not why they should move so as to get into contact with the particles of other substances in the first place. Thus Newton, as we have seen, found that the notion of 'hooked Atoms' begged the question.[2] Also, such properties as the spikiness of particles could not be more than speculation, and were (in modern terms) unverifiable. For the particles were by definition too small to be observed: only their effects appeared on the visible scale.

Mechanical explanations in terms of imponderable fluids such as light, heat, electricity, and phlogiston, consisting of particles which had properties of attraction and repulsion, were more popular in the eighteenth century. Yet in the end they did not provide entirely tenable explanations either, for no explanation was provided why a given kind of particle should repel its own kind, as was often required of imponderable fluids, but attract other kinds. Chemists were prepared to entertain such an explanation, but did not generally find it solved their own distinctively chemical kind of problem. They found it more helpful to follow another implication of the corpuscular philosophy, that is that each kind of particle had a fixed mass, and so came to think of imponderable fluids as ponderable and to try to identify them with particular substances. For instance, Kirwan and others tried to identify phlogiston with 'inflammable air' (hydrogen).[3] In other words, chemists developed their own autonomous concepts, and discarded or transcended the concepts of physics.

In deliberately using the words 'affinity' and 'attraction' to mean no more than 'tendency to combine', chemists clearly cut themselves off from any mathematical, mechanical, or philosophical explanation which the words might otherwise have implied. It was not that they believed such explanations were impossible. On the contrary, they generally seem to have hoped and expected that quantitative laws might one day emerge from the results of their labours. However, they must be laws which were appropriate to chemistry, and not based on the speculative and unhelpful systems of the physicists. The facts must be established first; then regular patterns must be found in the facts; and then some generalizations might emerge.

Often the point of using the words 'affinity' or 'chemical attraction' in the mid-eighteenth century was merely to give the feeling that something had been explained when in fact it had not. To say, for instance, that two substances combine because they have greater affinity with each other than with other substances, or because there is a stronger attraction between them, means no more than that they have been observed to combine with each other rather than with other substances. It is an explanation merely in the sense that it connects the tendency to combine shown by those two substances with all the other similar tendencies shown by other combinations of substances, and with the general view of such tendencies depicted in the affinity tables and in the various classifications of laws or types of affinity which will be described below. In other words, it is an explanation in the sense that it places a particular phenomenon in a class of phenomena that are already familiar, but not in the sense that it reveals causes or mechanisms or makes further predictions possible.

The practical value of the concepts of chemical affinity and attraction was partly, as we saw in the previous chapter, that with them in mind the chemist could draw up an affinity table and so produce a classification of substances. Even though simply applying the names 'affinity' or 'attraction' did not in itself accomplish much, and although it was quite possible to draw up tables of the order of reactivity of various substances without using those terms at all, as Geoffroy did, in fact it so happened that it was the concepts of affinity and attraction that his followers had in mind when they constructed their tables. From the tables chemists could then attempt to predict the results of previously unknown experiments which could be tested, at least in principle. Even if the tables were not very successful in predicting such results, they did at least bring about the observation of some new phenomena. A further step which has already been mentioned, taken in keeping with the eighteenth-century feeling for the importance of order and regularity, was to classify the different circumstances in which affinity or chemical attraction operated, and the modes in which it took effect. Although such classification was philosophically an end in itself, the discussion of how a particular reaction ought to be classified probably had some practical results, because it guided chemists in making new experimental observations.

Furthermore, however little it may achieve in the way of explanation, to attribute reactions to the operation of either affinity or attraction (it makes no difference which) does have one important implication. That is, affinity or attraction was taken to be a permanent and potentially measurable property of the reagents involved. Generally it was taken to be a property of the particles of which they were composed. However, although eighteenth-century chemists almost all believed that matter consisted of invisibly small particles, no experimental evidence was available before the time of Dalton that would have enabled a prudent experimentalist to discuss any properties of such particles individually as opposed to the properties of the matter which they constituted in bulk. The most that could be done was to try to measure or estimate the strength of the attraction between samples of visible size.

Yet the main value of the concept of chemical affinity or attraction in the later eighteenth century, as it turned out, was that it helped chemists to think of a chemical reaction as taking place between distinct substances that were unchanged in the course of the reaction, but would combine together or separate according to certain basic rules. The effects of the rules would, however, be affected by such factors as temperature, concentration, and the volatility of the possible products. Nowadays it would hardly occur to a chemist to think in any other way; but in the first quarter or even half of the eighteenth century the picture was by no means clear. Such thinking made possible the work of men like Richter and Berthollet.

Even apart from lines of thought which stemmed directly from the concepts of affinity or attraction, these concepts helped greatly towards providing eighteenth-century chemists with a mental picture or model of what chemical reactions were. With it in mind they were able to consider the properties and reactions of substances and interpret them, and so carry out further investigations that had more permanent results. To illustrate it, then, I shall in this chapter give some examples of attempts made by chemists in the latter half of the century to clear their thinking by classifying types of affinity or attraction, or by measuring it numerically, and then of the use made of the concept of affinity or attraction to provide what purported to be explanations. In doing so, I hope to convey something of the process by which chemistry developed its own ways of understanding and conceiving its subject-matter, independently of traditional physics, and so achieved autonomy.

The classification of types of affinity

Although with hindsight we may feel that the concepts of chemical attraction and affinity had such a limited explanatory power, the desire of eighteenth-century chemists to impose order and pattern on the disorder of natural phenomena could be satisfied in another way, by classifying types of affinity. Their classifications had as starting-point Geoffroy's very simple statement of the circumstances in which affinity operates, set out on p. 116 above. The first to develop this statement further seems to have been Macquer.

He devoted the second chapter of his *Elémens de chymie-théorique*, preceded only by a short introductory chapter on the Principles (Air, Water, Earth, Fire, and Phlogiston), to an 'Idée générale des rapports des différentes substances' (General idea of the relationships of the different substances). In it he gives an early form of his theory of affinity. He emphasizes that a knowledge of affinity is necessary to an understanding of chemical combination. His approach is empirical, and avoids any question of a mechanical explanation.

> All the experiments which have been made up to the present, and those which are still being made every day, combine to show that there is between different substances, whether principles or compounds, an agreement, relationship, affinity, or attraction if you prefer, which makes certain substances disposed to unite together, although they cannot contract any union with others. It is this effect, whatever may be the cause, which will enable us to give an account of all the phenomena which Chemistry furnishes, and to connect them together. The following shows in more detail how it is made up.[4]

It is significant that the phrase which has here been translated as 'to give an account' is in Macquer's French *rendre raison*—literally, 'to render reason', the word which stood particularly for the order and pattern which eighteenth-century philosophers sought to impose upon Nature.

Macquer then states seven propositions, which again are presented as generalizations from experience. That was the procedure recognized as correct, so that if he arrived at any theory it would be by induction and not by the imposition of an *a priori* system.

1. If one substance has affinity or *rapport* with another, they unite and form a compound.

2. Similar substances have affinity for each other, for example water for water and earth for earth.

3. Substances that unite lose a part of their properties, and compounds share in the properties of the principles forming them.

4. The simplest substances have the strongest affinities.

5. If a substance has no affinity for one of the two substances in a compound, but has a stronger affinity for the other substance in the compound than has the substance with which the latter is combined, then the new substance will form a new compound with it.

6. Sometimes a substance does not displace either of the two substances in a compound, but joins them to form a 'compound of three principles'. This happens when the new substance has an equal or almost equal affinity for both the substances in the original compound, or an affinity with one of them equal or almost equal to its own affinity for the other one.

7. If a substance has not sufficient affinity for one of the substances in a compound to decompose the compound by itself, it may be able to do so when combined with a fourth substance which has an affinity for the other substances in the original compound, and so may bring about a double decomposition and a double union.[5]

Macquer has assumed that affinity is a constant property of each substance, but has felt the need to account for situations where the simple rules of affinity do not apply, as in propositions (6) and (7), by supposing that special rules apply there. At any rate, the propositions are not really generalizations from experience, but rather represent a particular view of the nature of chemical reactions that is assumed rather than inferred, and then used to interpret the observed phenomena.

The first of the seven propositions is more a definition of affinity or *rapport*, or a statement of the circumstances in which one of those words will be used, than a description of experience. This and the last three propositions are developments of the general rule given by Geoffroy, and are accepted by most chemists of the later eighteenth and indeed of the early nineteenth centuries; but the second, third, and fourth are inherited from Stahl. The second proposition does not refer to the tendency of particles of the same substance to cohere together, which was carefully distinguished from chemical combination. Earth is a general category, derived from the Aristotelian element, and the notion of water is also that of an element or principle rather than of the familiar substance. Macquer is clearly thinking of the Stahlian theory that salts are composed of water and an earth, and that different salts (a category which at this time still includes acids and alkalis as well as neutral salts) combine together in virtue of containing the same earth. The theory was otherwise known through its mechanical interpretation as the theory of *latus*. The word 'affinity', then, is here used in a special Stahlian variety of the old sense, implying that substances tend to combine because of a similarity in their nature.

The fourth proposition is derived from the theory of Becher and Stahl that the simplest compounds, formed by the combination of primary particles, had weaker tendencies to combine than the primary principles, that the 'decompounds' formed by the combination of compounds were less reactive,and the 'surdecompounds' formed by the combination of decompounds less reactive still.[6] Macquer preferred to speak of 'compounds of the first degree', 'compounds of the second degree', and so on. The lower the degree of the compound, the stronger the affinity. A similar belief was held by Boyle and by Newton.[7]

The third proposition seems a little closer to being a generalization from experience. Nevertheless, what principles a compound was formed of was generally judged by finding what its properties were and assuming that it was made up of principles which possessed those properties; so that the proposition that 'compounds share in the properties of the principles forming them' is not based on observation but is simply a logical consequence of the means of determining what principles did form them. That is to say, it is a circular argument. It is true that loose compounds of compounds may retain some of the properties of the component compounds, and that is a generalization from experience; but that is not what Macquer means. The important point for him is that simple compounds show some of the properties of their component principles. This belief was only gradually replaced in the eighteenth century by the realization that the properties of compounds may be entirely different from those of the elements composing them. As we have seen, this realization came earlier to British than to Continental chemists.

Macquer illustrates the seven propositions in the rest of his book. Thus in the next chapter, 'Of Saline substances in general', he points out that salts are composed of earth and water (in which, as has just been explained, he follows Stahl) and that acids (his first instance of saline substances) have strong affinity with earths and water. Alkalis, which have a larger proportion of earth in them than acids, have less affinity for water. These relationships are not, of course, evident in Geoffroy's table, which Macquer reprints.

If the seven propositions which we have just discussed come rather nearer to an attempt at a set of laws of affinity, Macquer must have realized that it was a premature attempt. The first requirement was to find a regular pattern in the observed facts that were recorded in the affinity tables. As a step towards that, he and his partner tried to classify the different kinds of affinity that were observed. That would clearly be a considerable advance towards finding some regularities that could be used as a basis for laws.

On 7 April 1758 Macquer entered into partnership with Antoine

Baumé, then a young apothecary, to provide a public '*Cours de chymie et de pharmacie expérimentale et raisonné*'.[8] The course was delivered sixteen times, and ceased only in 1773 when Baumé was elected to the Academy. Baumé's *Manuel de chymie* of 1763 generally follows Macquer, and claims to give the basic doctrines of Becker, Staahl (*sic*), Boerhaave, and Macquer. At any rate, we can be fairly sure that Macquer would have approved of the contents of the book. Of affinities Baumé wrote:

> Monsieur Macquer has dealt with what relates to affinities in great detail in a particular memoir which provides the material of one of our lessons in Chemistry. He makes a methodical division of affinities, or rather a division of the different circumstances in which one meets affinities in the operations of Chemistry; although elsewhere he admits only a single type of affinity which is absolutely the same, and which he recognises to be derived from the same cause. Monsieur Macquer ranges under seven classes all Chemical affinities.[9]

There then follow seven classes, consisting of circumstances in which affinity is observed to operate, as Baumé points out, rather than of different classes of affinity.

1. *Affinity of aggregation*. This is the force that makes two bodies of the same kind tend towards each other, and adhere together when they are joined. It is exemplified by the adherence of two polished surfaces or the tendency of two drops of the same liquid to join.

2. *Simple affinity of combination*. Simple affinities of combination are those from which new compounds result, such as the solution of substances in acids. For example, if one puts white marble (i.e. calcium carbonate) into acid, it dissolves, and the resulting compound shares the properties of both the acid and the earth.

3. *Compound affinity*. Compound affinities are those of substances of different kinds that have an equal affinity together, the result of which is a blending with no decomposition. But the resulting blend has properties different from those of the different substances when separated. Baumé gives as an example the compound produced by melting together four parts each of lead and tin and adding two parts of mercury.

4. *Affinity of intermedium*. Such are the affinities of substances that can combine only by means of another that has affinity with both of them. For instance, marble will not on its own combine with water; but if nitric acid is added it acts as an intermedium, and enables the marble and water to combine. (This is rather like what Henkel had called *appropriatio* and Weigel was later to call *affinitas appropriata*.)

5. *Affinity of decomposition*. Affinities of decomposition are those from which result decompositions and new combinations. For instance, if fixed alkali (potassium or sodium hydroxide) is added to a solution of

marble in nitric acid, the alkali combines with the nitric acid and the earth of marble (calcium hydroxide) is precipitated.

6. *Reciprocal affinity*. Reciprocal affinities are those from which reciprocal decompositions result, and correspond with what would now be called reversible reactions. For instance, nitre (potassium nitrate) is decomposed by vitriolic (sulphuric) acid, because the acid disengages the nitric acid and combines with the alkali, forming 'vitriolated tartar' (i.e. potassium sulphate). However, if nitric acid is added to vitriolated tartar in its turn it disengages the vitriolic acid and reforms the nitre. As has already been pointed out, the term 'reciprocal affinity' seems first to have been used in print by Marherr in 1762; though it is just possible that he may have heard of its being used by Macquer or Baumé in their public course of lectures. It is hard to see any good reason for refusing to admit that cases of reciprocal affinity are cases where the affinity between substances varies and is not constant.

7. *Double affinity*. Double affinities of four substances are those from which result two decompositions and two new combinations by reciprocal exchanges—in other words what would now be called double decompositions. Baumé gives several examples, the first being the decomposition of vitriolated tartar (potassium sulphate) and Glauber's salt (sodium sulphate) by all solutions of metals in nitric acid.[10]

It is clear that Macquer and Baumé have moved on from the idea that affinity implies similarity of composition, and are using the word to mean simply tendency to combine. They have also tacitly abandoned Stahl's mechanical theory of 'latus'. Nevertheless, it is still implied in this statement that, at any rate for affinities of the second type, the compound has properties that are a blend of the properties of the constituents, and not an entirely new set of properties. For affinities of the third type, on the other hand, there is the important new point that a compound formed by their means will have properties that are not just a blend of those of the constituents. It is not quite clear whether the particular property of affinity is modified when a substance is combined or not.

The third, fourth, and fifth types of affinity given by Baumé are developments of the last three of the seven propositions stated earlier by Macquer in his *Elémens*.

In the first edition of his *Dictionnaire de chymie*, published anonymously in 1766, Macquer expressed the seven types of affinity in a slightly different way.[11] In particular, he distinguishes the first two, as types of Simple Affinity, that is affinity between two substances, or parts of the same substance, from the rest, which are types of Complex Affinity, involving more than two substances, and are varieties of the second type. The first two types are (1) Affinity of Aggregation (as in Baumé's *Manuel*), by which particles of the same kind are united, and (2) Affinity of Composition, by which particles of different kinds are united to form a compound. The other five types are essentially the same as those given in Baumé's *Manuel*. However, Macquer does not give a specific name in his description of the third type, which he later refers to as Affinity of Three Principles (Baumé's Compound Affinity), or to the fifth type, which in Baumé is Affinity of Decomposition. Also, Macquer makes it clear that what he says about the Affinity of Three Principles applies *mutatis mutandis* to Double Affinity, where four principles are concerned. In short, Macquer makes the connections between the various types rather clearer than does Baumé.

In his *Chymie expérimentale et raisonée* of 1773, a much more elaborate work than the *Manuel de chymie*, Baumé presents the classification of affinities again with slight differences.[12] He begins by saying that although chemical affinities are all due to attraction they present different effects according to the state of the substances involved. This reservation chimes in with the comment frequently made about affinity tables that they do not take varying conditions into account. The first case considered is a new addition to the list, 'Affinity of adherence or cohesion'. This evidently means the force which causes two different objects to cling together in mechanical rather than chemical combination, and is distinct from 'Affinity of aggregation', now second on the list, which refers to the cohesion of the parts of a homogeneous piece of matter. The other six cases follow as before in the *Manuel de chymie*, except that 'Simple Affinity of Combination' has become 'Compound Affinity of Two Bodies', as opposed to the following case, which is 'Compound Affinity of Three Bodies'. 'Affinity of Decomposition' has become 'Affinity of three bodies, from which results a decomposition and a new combination which takes place at the same time'; and 'Double Affinity' has become 'Affinity of four bodies, or Double Affinity, from which result two decompositions and two new combinations'. The new names for the types of affinity make the meaning clearer, even though they are more cumbersome.

It is made even plainer in this latest version of the classification that the affinity of aggregation between particles of the same substance is quite distinct from the affinity between particles of different substances that produces chemical combination. The distinction becomes natural as soon as a systematic classification of affinities replaces the Stahlian idea that affinity between substances is due to their having constituents in common, though it is itself merely a development of the distinction between a mechanical mixture and a chemical compound. The motive for emphasizing the distinction may have been a desire to keep the subject-matter of a chemistry rapidly approaching autonomy, which required special techniques and a special approach for its understanding, clearly separate from the subject-matter of physics. At any rate it was a distinction which made it difficult for Dalton and others of his time to accept Avogadro's suggestion that in elementary gases atoms of the same substance could be united chemically to form molecules.

Clearly the gradual changes in interpretation of chemical reactions expressed in the various stages of these classifications reflect much the same developments as the changes in affinity tables recorded in the previous chapter.

Later writers in the eighteenth century often used classifications similar to Macquer's and Baumé's, though on the whole they tended to simplify them.[13] The best-known example is Bergman's classification of the various kinds of contiguous attraction, by which he meant the attraction between bodies close together near the earth's surface.[14] It is contrasted with the attraction operating at long ranges, to which Newton's law applies, since at short ranges (as Buffon had pointed out) particles cannot be regarded as having negligible size compared with the distance between them (as heavenly bodies may be), and the shape of each particle therefore produces a total effect which appears not to be in accordance with Newton's law.

The union of homogeneous bodies, when the only change is an increase of mass and the nature of the substance is the same, is Bergman's instance of the *attraction of aggregation*. On the other hand, heterogeneous substances in forming combinations are affected by differences of quality rather than quantity. This involves *attraction of composition*. When this unites two or more substances, it is called *attraction of solution* or *of fusion*, according to whether the union takes place in the wet or the dry way. When it takes place between three substances, of which two unite and the third is excluded, it is said to be a *single elective attraction*; when it takes place between two compounds, each consisting of only two proximate principles, which are exchanged in the reaction, it is called *double attraction*. It is this last species which Bergman represents in his diagrams of double decompositions. After him authors of affinity tables, as we have seen, tried to show single, double, and sometimes more complex affinities or attractions separately.

An example of an even more simplified form of the classification is that of Gren. He brings chemical affinity or elective attraction under three heads:[15]

1. *Affinity of composition*, or mixing affinity (*affinitas synthetica*), whereby two or more heterogeneous substances unite to constitute a new and thoroughly homogeneous whole. Instances are the combination of water and salt, of alcohol and resin, of sulphuric acid and 'argil' (i.e. alumina), of nitric acid and potash (potassium hydroxide), of silver and sulphur, of silver and gold, and of silver, gold, and copper. It strikes the modern reader at once that Gren does not distinguish between solution, alloying, and chemical combination in the strict sense. However, it was

normal in his time to make no distinction, and it does not affect his general argument. Affinity of composition includes *appropriation* or *intermediate affinity*, through which two substances showing no affinity are united by a third. This corresponds with Macquer's Affinity of Intermediums, which Bergman had not included in his classification. Gren's instances of this sub-species are the affinity of fatty oil, water, and alkali, and of sulphur, water, and alkali. *Preparing affinity*, he adds, is also a case of affinity of composition. By this he means cases where one substance prepares another for union with a third for which it would otherwise show no affinity.

2. *Simple affinity* (*affinitas electiva simplex, analytica cum synthesi simplici*, or simple elective affinity, separative together with simple mixing) takes place when two heterogeneous substances, united to one another in a homogeneous whole, are parted by adding a third, which attracts one of the two united substances more forcibly than they attract each other, and thus one of the two is separated from the other. Gren shows three examples schematically:

He adds:

> Hitherto it has in vain been attempted to find out a general law by which these affinities act; for there has not yet been collected a sufficient number of facts to discover that law. Yet towards a general view of the experiments belonging to this subject, we are greatly assisted by the *Tables of simple affinities*, where the various substances are arranged in succession, according to the greater or less elective attraction which each of them has for any particular substance.[16]

This passage sums up very clearly the intentions behind the various classifications of affinity.

3. Gren's third species is double affinity, whereby more than one new combination is caused, or two combined substances are separated on the addition of two others (no matter whether combined or uncombined) by virtue of their respective attractions, so that two new combinations arise, though only one separation may take place. Double affinity often causes a decomposition which the simple affinity is unable to effect. Again, Gren gives three examples schematically.

Although it is not explicitly stated, it appears from the arrangement of his sections that Gren is unusual in regarding chemical affinity as a variety of the more general attraction of cohesion. It would therefore have been difficult for him to see any distinction between solution, the formation of alloys, and chemical combination, even if such a distinction had been normal at the time. Nevertheless, by reducing the main types of chemical affinity to three he does succeed in making the basis of the classification stand out more clearly.

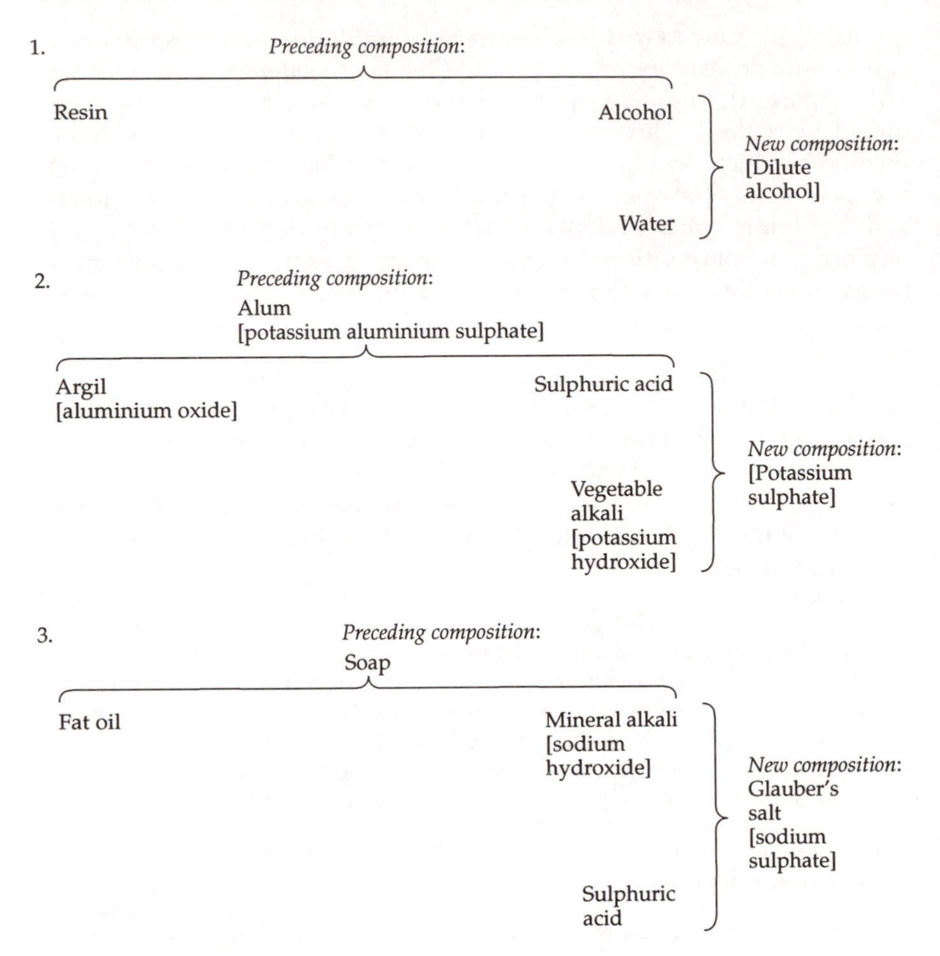

Fig. 5.1 Three diagrams by Gren illustrating the operation of *simple elective affinity*, when two heterogeneous substances united in a homogeneous whole are separated by a third substance that attracts one of the two previously united substances more forcibly than they attract each other. From F. A. C. Gren, *Grundriss der Chemie*, 2 vols, Halle, 1796–7.

However, the classification is very little help in the attempt to find a general pattern from which it would be possible to proceed to a set of laws of affinity by which precise predictions could be made. Gren does, indeed, do his best by stating as a general law: 'Every chemical decomposition is performed by the assistance of affinities, and two substances united into one homogeneous compound can never be separated, unless one of them undergoes a new combination.'[17]

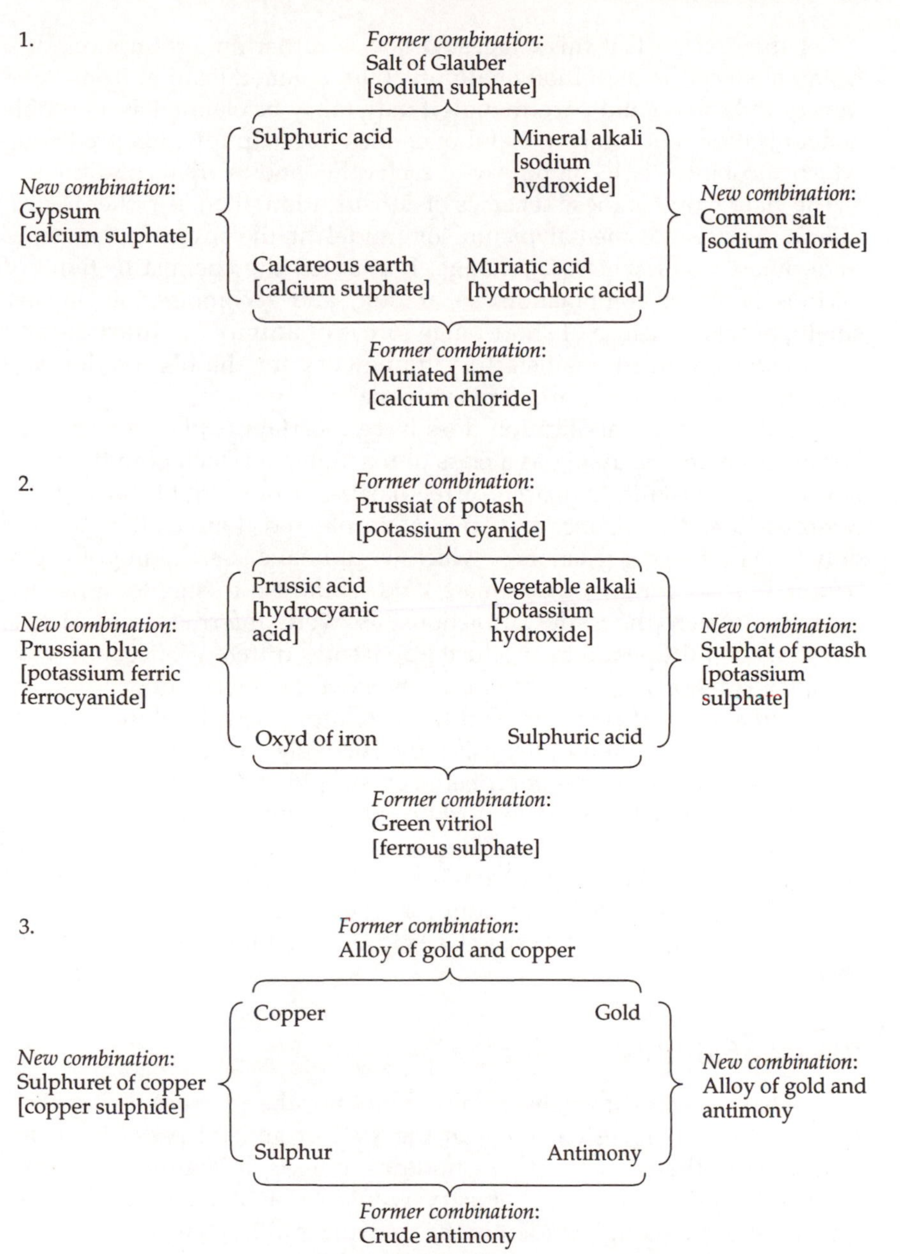

Fig. 5.2 Three diagrams by Gren illustrating *double affinity*, whereby a combination of two substances is decomposed by the addition of two other substances (usually also previously in combination with each other), with resultant dissolution of the original compound (or compounds) and production of two new compounds by recombination. From F. A. C. Gren, *Grundriss der Chemie*, 2 vols, Halle, 1796–7.

Yet the first part of this suggested law is rather an assumption, or a restatement of the definition of affinity, than a generalization from experience; and the second part, though it may fairly be claimed as a testable generalization, and quite a useful one, does not help towards predicting which substances will combine with each other and in what conditions.

The chief value of these schemes of classification, then, is rather to provide a serviceable mental picture or model of the invisible processes underlying visible chemical change. It enabled the chemist to think of various types of combination, separation, and recombination of very small particles because of short-range forces of affinity or attraction; and this model provided a satisfactory framework for the observation and description of many important phenomena.

In addition, the classification does have a certain explanatory power. For instance, in discussing as a class of reactions in which affinities seem not to be constant 'Apparent irregularities from double attraction', Bergman uses the distinction between simple and double attraction not only to classify such reactions—which would have been a kind of explanation, but to him not an adequate kind—but also to suggest a mechanism by which the same attractions as were inferred from simple reactions could operate to produce apparently different effects in more complex reactions.[18] The mechanism, however, does not depend on the sizes and shapes of particles, as if he were a physicist thinking in terms of the mechanical philosophy: it is a distinctively chemical one. It might be expected that differentiation between simple and double or complex attractions would have led to further experiments by which the hypothesis that some reaction was due to the operation of complex rather than simple attraction might have been verified. However, the question whether a reaction should be counted as simple or complex is argued from existing experimental evidence, and does not lead to further experiments.

'Laws' of affinity

As we have seen, neither the affinity tables nor the schemes of classification of affinity suggested much in the way of general patterns or any indication of the working of general laws. Indeed, it is difficult to imagine what kind of laws of chemistry would be analogous to Newton's laws of motion, though evidently the unexpressed hope was to find laws which would enable a chemist to predict exactly whether one substance would combine with another or not. The difficulty was not a lack of facts, but that there were too many of them that did not seem to fit in with any ordered scheme. Yet in spite of difficulties, the aim of inferring general laws from regularities in the observed data, according to the best principles of enlightened philosophy, was very attractive. It is not unnatural that of the few who attempted it, two, Fourcroy and Guyton de Morveau, also tried to quantify the force of chemical attraction and affinity, which of course was the correct Newtonian procedure, though they could not reach any satisfactory conclusion.

The starting-point for laws of affinity was not so much generalization from the facts as Geoffroy's original proposition, stated in his paper of 1718:

> Every time that two substances which have some disposition to join with each other happen to be united together, if there supervenes a third which has more relationship with one of the others, it unites with it and makes it release its hold on the other.[19]

That it is a proper interpretation of such a reaction to attribute it to the difference in closeness of relationship (*rapport*) or affinity, or in the force of attraction, is obviously no more than an untested assumption. Also, it is an assumption which it is difficult to reconcile with the apparent variations in the order of affinity shown by such phenomena as reversible reactions. Nevertheless, the desire for regularity was strong enough to override problems of that sort.

Geoffroy's ideas are extended slightly by the author of the *Nouveau cours de chymie*, first published in 1723 and again in 1737. However, the most notable attempt at a set of general laws of affinity was that published by Fourcroy in 1782. They obviously owe a good deal to the classifications of affinity published by Macquer and Baumé. Born in 1755, Fourcroy was in 1782 at the start of his career, and the *Leçons élémentaires d'histoire naturelle et de chimie* which includes the laws was his first book. It contains a 'Dissertation sur les affinités chimiques', in which affinity of aggregation is distinguished from affinity of composition. The latter is defined as follows:

> When two bodies of different nature tend to unite, in that case they combine in virtue of a force different from that which we have examined above [i.e. affinity of aggregation uniting bodies of the same nature], and to this force is given the name of *affinity of composition* or of combination.

After emphasizing that knowledge of this force should be the guide of both the practical worker (*l'artiste*) and the savant, Fourcroy continues: 'These laws, founded on a great number of positive experiments, can be reduced to ten.' He then sets out the laws themselves, each headed 'Law of the Affinity of Composition' and accompanied by an explanation.

1. The affinity of composition takes place only between bodies of different nature.

2. The affinity of composition takes place between bodies only at the point of contact.

3. The affinity of composition takes place only between small bodies.

4. The affinity of composition can take place between several bodies.

5. For the affinity of composition to take place between two bodies, it is necessary that one of them, at least, should be fluid.

6. The affinity of composition is in inverse ratio to the affinity of aggregation.

7. When two or more bodies unite by affinity of composition, their temperature changes at the instant of their union.

8. Two or more bodies which are united by affinity of composition form an entity which has properties which are new and very different from those which the separate bodies had before uniting.

9. The affinity of composition is measured by the difficulty which is experienced in separating the combination formed between two or more bodies.

10. All bodies have not the same degree of affinity between them, and it is possible with the aid of observation to determine the rank or degree of this force that exists between the different bodies in nature.[20]

Most of these laws are restatements of generally accepted principles. In his discussion of the first, which follows naturally from the distinction between affinity of aggregation and affinity of composition, he makes it clear that he is rejecting the view held by Stahl and others that substances combined only in virtue of a *rapport* or a resemblance between their properties. (This is not, of course, a reference to Geoffroy, in spite of the word *rapport*, since he was not one of those who held this view.) The third law makes it clear that affinity is a property of the insensibly small particles of which matter was assumed to consist, though the word 'atoms' is not normally used at this period. The seventh law, stating the change of temperature at the instant of chemical union, probably shows the influence of Lavoisier. His memoir on combustion, arguing that the matter of fire escapes when a body burns, was read in 1777 and published in 1780. It may well have drawn Fourcroy's attention to the change of temperature associated with chemical change, even though he does not accept Lavoisier's explanation. Lavoisier's work with Laplace on heat was carried out probably in 1783–4.[21]

The most interesting of the ten laws is the eighth, stating that a compound has different properties from its constituents. It goes rather further than the older view expressed by Macquer in 1749 that the properties of the constituents are only modified, and the change is an assertion of the growing autonomy and self-confidence of chemistry. The law stems from the practical observation of chemists rather than from the theoretical picture of mechanical philosophers. The ninth law, on measuring the affinity of composition by the difficulty of separating bodies, no doubt reflects the work of Guyton de Morveau, which will be described below.

Fourcroy later revised these laws, the most notable change being that he renamed them 'laws of attraction of composition' instead of 'affinity of composition'. This change marks the end of the process of emptying the words 'affinity' and 'attraction' of any implications of speculative or mechanical explanations. It is probably due to the general acceptance as authoritative of Bergman's *Dissertation on elective attractions*, though of course it had already been published in its original form in 1775, and Fourcroy referred to Bergman in his 1782 edition.

The main changes in the revised form of Fourcroy's laws are that he omits the second and sixth of the 1782 laws, and later includes two new ones.[22] The second of the 1782 laws is not very helpful, but seems unobjectionable—perhaps in omitting it Fourcroy wished to leave open the possibility that attraction operated at a distance, however short, or at least not to enter into the controversy on this point. The sixth of the 1782 laws, that affinity is inversely proportional to aggregation, is presumably dropped because it is misleading. Fourcroy had read too much into the obvious point that chemical combination is resisted by the cohesion of the substances which tend to combine.

The two new laws which are included in 1800 are:

9. The attraction of composition follows the inverted ratio of the saturation of bodies by each other.

10. Decomposition may take place between two compounds, which do not mutually decompose each other by double elective attraction, if the attraction of two of the principles for a third be stronger than that which unites this third to one of the two first, though, at the very moment, the act of union between these two first principles does not yet exist. (This is called 'predisposing attraction'.)

The former of these implies acceptance of Kirwan's views, which had been published only in 1781, and were discussed by Guyton de Morveau in his *Encyclopédie méthodique* in 1786. Fourcroy himself in his *Mémoires et observations de chimie*, published in 1784, had used an arbitrary method of allocating numerical values for the strength of affinities. Presumably he had come to accept Kirwan's views in the interval between 1784 and the issue of the two new laws in 1800. Fourcroy mentions, in his commentary on the law that the attraction of composition is measured by the force required to separate the components, that the speed of combination does not give a means of measurement. That is a reference to the work of Wenzel, which had also been discussed by Guyton.

The second of the two laws which are new in Fourcroy's 1800 version, numbered 10, is a development of the ideas of Macquer, but no doubt owes more to Bergman. Nevertheless, its inclusion is probably related to Fourcroy's acceptance, in a paper added at the end of his *Mémoires et observations*, that in his original allocation of numerical values for affinities he had neglected what Kirwan called the quiescent affinities between two of the four components in a double decomposition—that is, those which resist decomposition.

Meanwhile, Guyton had produced in his *Encyclopédie méthodique* a rather simpler set of six laws.[23]

1. There is no chemical union at all, if one of the bodies is not fluid enough for its molecules to obey the affinity which carried them from proximity to contact.

2. Affinity occurs only between the smallest integrant molecules of bodies.

3. One should not assume from the affinity of one substance with another that the compound of these substances will have affinity with one or the other in excess.

4. The affinity of composition is effective only in so far as it overcomes the affinity of aggregation.

5. Two or more bodies that unite by affinity of composition form an entity which has new properties distinct from those that belong to each of the bodies before combination.

6. Affinities are affected by temperature, which makes their action either slow or fast, or ineffective or effective.

Fourcroy and Guyton were associated in several pieces of work, particularly in the new system of nomenclature of which they were joint authors together with Berthollet and Lavoisier. It will be seen that Guyton has omitted the first, fourth, and tenth of Fourcroy's original laws, which indeed are rather obvious generalizations, and has nothing corresponding to the second of the laws which Fourcroy later added to his revised list. Also, Guyton has dealt with the measurement of affinity elsewhere in the article in which his laws occur, so that he has no law corresponding to the ninth of Fourcroy's original list or the former of the two additional laws in his revised list. Numbers 1, 2, 4, and 5 of Guyton's laws correspond with numbers 5, 3, 6, and 8 of Fourcroy's earlier list, and the first of Guyton's laws also covers the second of Fourcroy's original laws, that affinity applies only to substances in contact. However, Guyton's fourth law, that affinity of composition is effective only in so far as it overcomes the affinity of aggregation, is much less specific than Fourcroy's corresponding law that affinity of composition is actually in inverse proportion to affinity of aggregation, and is presumably connected with the second thoughts which induced Fourcroy to leave this proposition out of his revised list. Also, Guyton's sixth law, that affinities are affected by temperature, is almost the converse of the law that Fourcroy has in both his lists that there is a temperature change at the instant of chemical union.

Other examples of such classifications of affinity or laws of affinity could be mentioned; but those which have been quoted give an adequate representation. It is evident that none of the suggested sets of laws comes anywhere near enabling chemists to predict whether two substances would combine or not, nor are they based on regularities found in affinity tables; but at least Fourcroy and Guyton had done their best according to the principles of eighteenth-century philosophy. Their laws are rather descriptions of the conditions in which affinity or attraction operates than laws governing their operation. Yet they do state as much in the way of generalization as was possible in the light of contemporary understanding, and they do, as I have suggested, represent a serviceable model of chemical reactions.

With hindsight, we may say that the attempt to arrive at general laws through study of the tendency of substances to combine, whether it was called affinity or attraction, was doomed to failure, because far too few of the relevant factors were understood. The way ahead turned out to be quite different, through measurement of the weights of substances that combined together, atomic weights, valency, and the periodic table. Nevertheless, eighteenth-century chemists did try to make quantitative measurements or estimates of the force of affinity.

Attempts at quantification

The measurement of the actual magnitude of the forces involved was one direction in which the study of chemical affinity and attraction did suggest new experiments. This was the natural step towards quantification and the eventual discovery of mathematical laws. There were three main types of method:

1. Allocating numerical values to the affinities between the various substances in an affinity table by trial and error, starting with values chosen at random, so that the substances in any reaction that had the affinity represented by the largest number were the ones that actually combined.

2. Measuring the force of attraction between substances mechanically.

3. Measuring the quantities of two substances that combined together at saturation point, or the speed with which substances combined, and so from a measurable property of chemical reactions calculating numerical values for the affinities concerned on some assumption about the relationship between affinity and the factor that had been measured.

Such methods obviously imply that the force of chemical attraction cannot be directly inferred from the laws of gravitational attraction, though they are not inconsistent with Buffon's view that chemical attraction is simply gravitational attraction modified by the shapes of the particles concerned. In that case, from a knowledge of the apparent laws of chemical attraction, and of the fact that these could be reduced to complex instances of the law of gravitation, the shapes of the particles might in principle be deduced. However, none of the chemists who used these methods discussed that point. In practice, the question whether chemical attraction was the same as gravitation or not did not occur to them as a realistic question to ask at all.

The first method, that of trying arbitrary numbers to represent affinities, no doubt originates with Black's device of putting numbers to show the relative strength of the forces involved in a double decomposition. It is almost certain that Black did not intend these numbers to be taken seriously—they were used merely to illustrate the principles of double decomposition for pupils at an elementary stage of learning chemistry. The device was, however, suggestive.

In his *Elements of natural philosophy* Elliot, who had studied Black's lectures, reprinted the table of sixty-four diagrams of double decompositions from Bergman's *Dissertation on elective attractions*, and added a sixty-fifth in which figures were inserted to show the actual magnitude of the affinities.[24]

This means that when (in modern nomenclature) potassium sulphate and silver nitrate, which are symbolized on the right and left sides of the diagram respectively, react together, silver sulphate is precipitated and potassium nitrate (nitre) remains in solution. These products are shown at the bottom and at the top of the diagram respectively, and the four components are shown at the four corners. The figures mean that the attraction of potassium hydroxide (pure vegetable alkali) for nitric acid is 8, that of potassium hydroxide for sulphuric acid is 9, that of sulphuric acid for silver oxide (the calx of silver) is 4, and that of the silver oxide for nitric acid is 2. Calx of silver is equivalent to what would now be regarded as the oxide of the element silver, but in the phlogiston theory was itself regarded as the elementary substance of which metallic silver was a compound. Since (8+4) is greater than (9+2), the double decomposition takes place even though the attraction of potash for sulphuric acid is the strongest of the four on its own. The figures were not found by experiment, but by assigning arbitrary numbers, as Black had done, for one reaction and adjusting them by trial and error to fit other reactions (see p. 146). However, no one set of numbers could be made to fit all reactions, and so the method was bound to fail unless other factors such as temperature and concentration were taken into account. Even then, there would have been many anomalies. It is remarkable that Elliot

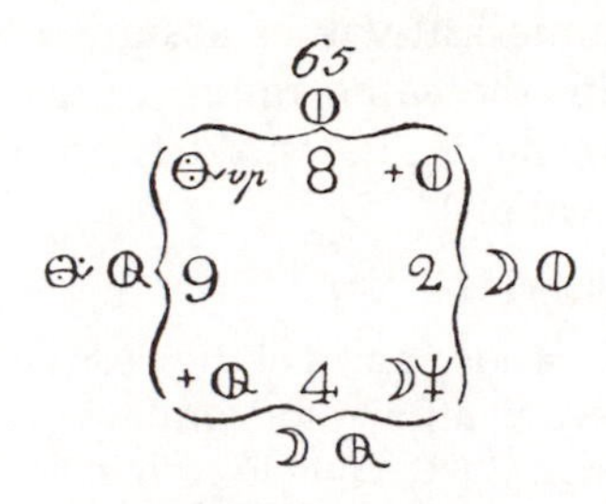

Fig. 5.3 The sixty-fifth diagram from J. Elliot's table headed 'Bergman's second table': an additional diagram by Elliot, with figures showing the relative magnitude of the affinities. This example of double affinity shows (in modern terminology) that when potassium sulphate (right-hand side, outside braces) and silver nitrate (left-hand side, outside braces) react together silver sulphate (bottom, outside braces) is precipitated and potassium nitrate (nitre, top, outside braces) remains in solution. The figures (arrived at by trial and error) constitute a claim that the attraction of potassium hydroxide (pure vegetable alkali, top left inside braces) for nitric acid (top right inside braces) can be represented as 8, that of potassium hydroxide (top left inside braces) for sulphuric acid (bottom left inside braces) as 9, that of sulphuric acid (bottom left inside braces) for silver oxide (the calx of silver, bottom right inside braces) as 4, and that of silver oxide (bottom right inside braces) for nitric acid (top right inside braces) as 2; and that therefore, since the sum of 8 + 4 is greater than that of 9 + 2, the double decomposition and recombination will take place even though the attraction of potassium hydroxide for sulphuric acid carries the highest value of any of the four attractions taken individually. Note that in the phlogiston theory calx of silver (silver oxide) was regarded as itself an elementary substance of which metallic silver was a compound. From J. Elliot, *Elements of the branches of philosophy connected with medicine*, London, 1786 (1782).

misses out the symbol for water that Bergman puts in the middle of his diagrams for reactions in the wet way.

A similar attempt was described by Fourcroy in a memoir which he wrote in November 1783 and read to the Académie Royale des Sciences in 1784.[25] Its purpose was to explain the workings of 'affinité double', by which two substances in combination could decompose a compound of two other substances which neither of them could decompose on its own. 'I thought it right to make known in more detail the method which I employ to conceive and to determine with the greatest exactitude the cause of this sport of affinities.'[26] Fourcroy adds that nobody before him had adopted the method which he proposes to explain double affinities, so presumably he is unaware of Elliot's work. However, he obviously owes a greater debt to Bergman, whose contribution he does acknowledge.

Fourcroy uses selenite (calcium sulphate) as an example of his method. First, he supposes that in selenite the vitriolic (sulphuric) acid adheres to the chalk (i.e. the calx of oxide of calcium) with a force equal to four. At

Table 5.1. Fourcroy's numerical values assigned to the affinities of four acids for combining with the same range of six substances (Fourcroy, *Mémoires et observations de chimie*, Paris, 1784, pp. 314–15 (original values) and 433–4 (later revised values)

Substances combined with Acids

The acids named at the heads of each of cols. 1–4 have an affinity for combining with the substances listed at left equal to (original values in clear and revised values in brackets):

	Col. 1 **Vitrolic acid**	Col.2 **Nitrous acid**	Col. 3 **Marine acid**	Col. 4 **Carbonic acid**
Fixed vegetable alkali	8(8)	7(7)	6(6)	3(3)
Fixed mineral alkali	7(7)	6(6)	5(5)	2(2)
Quicklime	6(6)	5(4)	4(3)	1(1)
Volatile alkali	5(4)	4(3)	3(2)	$\frac{3}{4}(\frac{3}{4})$
Magnesia	$4(\frac{1}{2})$	3(2)	2(1)	$\frac{1}{3}(\frac{1}{3})$
Alumina*	3(2)	2(1)	$1(\frac{1}{2})$	$\frac{1}{4}(\frac{1}{4})$

*The word 'alumina' represents Foucroy's *argile*.

this stage, the choice of number is quite arbitrary. Since ammonia cannot separate the acid from the chalk, he concludes that the acid and ammonia have a tendency to combine which is equal to three. Similarly, carbonic acid cannot separate the chalk from the vitriolic acids, because it has for this 'salty-earthy substance' only an affinity which Fourcroy compares with the number three. However, in that case ammonia and carbonic acid acting together have a total action equal to six, which is greater than four, and so will separate the constituents of the selenite.[27]

The obstacle in applying this method to all cases is that it would require a different number to be allotted for every entry in the affinity table, not only for the substance at the head of each column. Fourcroy does not try to do that, but gives four columns as examples. These are translated in Table 5.1 with both the original values which he gives in this and the revised values which, as we shall see, he gives in a later version.[28]

In his original memoir Fourcroy goes on to say that this attempt could be extended to all substances that have some affinity with each other, but would demand more exact information than has hitherto been acquired. He then illustrates the use of the numbers by inserting them in Bergman's diagrams of double decomposition, though he shows the substances taking part in words and not, as Bergman had done, in symbols.[29] However, he does put in the symbol for water (an inverted triangle) in the middle of all these diagrams except for the first, to show that the reactions are in the wet way. (It is left out of the first, presumably by an over-

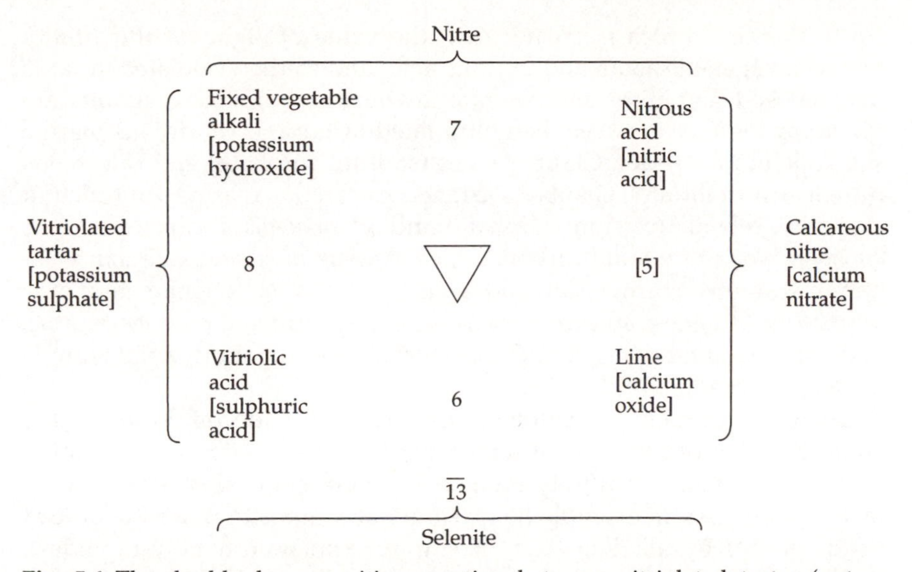

Fig. 5.4 The double-decomposition reaction between vitriolated tartar (potassium sulphate) and calcareous nitre (calcium nitrate) as first illustrated, with the original numerical values given for the affinities involved, in A. F. de Fourcroy, *Mémoires et observations de chimie*, Paris, 1784, p. 316. The bracketed figure '5'at right for the affinity between the substances shown on the right (nitrous—i.e. nitric-acid, top right, and lime, bottom right) did not form part of the original diagram, though it can be derived from the table of numerical values for the affinities of four acids published along with it (ibid., pp. 314-15 = Table 5.1 (original values) of the present work); but including it would have brought Fourcroy's sums of affinity values into equilibrium in this case (7 + 6 = 13: 8 + 5 =13), thus failing to provide any reason for the double decomposition.

It was to cope with criticisms of problems of this nature that Fourcroy produced his later, revised table of values for the affinities (ibid., pp. 433-4), which in this case would reduce the value for the right-hand-side affinity between nitrous (nitric) acid and lime to 4, thereby decreasing the value for the quiescent (horizontal) affinities to 8 + 4 = 12, while leaving that for the divellent (vertical) affinities as 7 + 6 = 13, thus again providing a set of figures that could account for the double decomposition.

sight.) Nevertheless, he does not consider whether some numerical value should be put on the effect which the fact that the reagents are dissolved in water might have on the course of the reactions.

The first example is the reaction between vitriolated tartar (potassium sulphate) and calcareous nitre (calcium nitrate) (Fig. 5.4). The values for the affinities between vegetable alkali (potassium hydroxide) and nitrous (i.e. nitric in modern terms) acid, vegetable alkali and vitriolic (sulphuric) acid, and lime (calcium oxide) and vitriolic acid are taken from the table and inserted in the diagram. The values for the affinity between vegetable alkali and nitrous acid and between lime and vitriolic acid add up to thirteen, which is greater than the value of eight for the affinity between vegetable alkali and vitriolic acid, and so the vitriolated tartar is decomposed and nitre and selenite are formed. Similar diagrams are given for the reactions between vitriolated tartar and calcareous marine salt (calcium chloride), Glauber's salt (sodium sulphate) and calcareous nitre (calcium nitrate), Glauber's salt and calcareous marine salt (calcium chloride), selenite (calcium sulphate) and *sel ammoniacal craieux ou Craie ammoniacale* (ammonium carbonate), calcareous nitre and *craie ammoniacale*, calcareous marine salt and *craie ammoniacale*, selenite and *craie magnésiene* (magnesium carbonate), calcareous nitre and *craie magnésiene*, and calcareous marine salt and *craie magnésiene*—in short, a fair sample of elementary mineral chemistry.

Lastly in this memoir, Fourcroy emphasizes that decompositions that could be produced by one substance acting alone should not be counted as instances of double affinity even if two intervening substances were in fact involved. As an example he gives the decomposition of nitre of lead (lead nitrate) by alkaline liver of sulphur (potassium polysulphides), yielding nitre (potassium nitrate) and galena (lead sulphide), in which the affinity of seven between the nitrous acid and the vegetable alkali (potassium hydroxide) alone would overcome the affinity of only one between the calx (oxide) of lead and the nitrous acid.

An obvious criticism of this exposition is that Fourcroy entirely neglects the affinity between the two substances on the right of the diagram. In the example of the reaction between vitriolated tartar and calcareous nitre, for which the diagram has been shown above, it would be logical to write in the value of the affinity between nitrous acid and lime, which from the table is five. The sum of the affinities between the two original pairs of substances would then work out as thirteen (eight plus five). That is the same as the sum of the affinities between the two substances which are formed as the result of the double decomposition, and Fourcroy's point would be lost.

In a paper added at the end of his *Mémoires et observations* Fourcroy acknowledged this flaw in his original exposition. Possibly it had been pointed out to him when he read the first paper to the members of the *Académie Royale*. He therefore adopted Kirwan's distinction between the affinities tending to keep the substances in their original pairings, the 'quiescent affinities', and the affinities tending to decompose the original pairings and produce new ones, the 'divellent affinities'.[30] The quiescent affinities would be those between substances which were next to each other vertically in the diagram of the double decomposition, and the divellent affinities between those which were next to each other horizontally. He now admitted that the quiescent affinity between the two substances at the right of the diagram should be included. Consequently, he had to amend some of the values given for the affinities in the original memoir. The amended values have been shown in Table 5.1 alongside the original values. If the new value for the affinity of nitrous acid and lime is used in the diagram for the reaction between vitriolated tartar and calcareous nitre, the sum of the quiescent affinities now becomes twelve, and the sum of the divellent affinities remains thirteen; so the result of the reaction (a double decomposition) is correctly predicted. In modern terms,

Potassium sulphate + calcium nitrate =
potassium nitrate + calcium sulphate.

Fourcroy proceeds to insert the new values in the other nine diagrams, which now also predict the correct results.

He does not in any way imply that the numbers which he gives for affinities have any physical meaning, or absolute validity, and he gives no units for them. Nor does he mention the possibility of further experiments to test the validity of the numbers in other circumstances. Nevertheless, he does imply that the revised numbers were to be taken at least fairly seriously, for although he says that his first intention was to make the explanation of double affinities clearer than it was in books of chemistry, for the benefit of beginners in the subject, he goes on to say that since the original memoir was written he has given it greater precision and has taken account of necessary additions and corrections. However, he must have been well aware that less arbitrary methods of quantifying the strength of affinity, based on some theory, were possible. He knew the work of Guyton and Kirwan, to which indeed he referred in the ninth law of both his original and his revised sets of laws of affinity. Possibly, although he does not say so, he had the reasonable intention of finding numerical values which fitted the facts without drawing on any theoretical preconceptions, and then trying to find a pattern in the numbers which would lead to a stronger set of laws.

Yet it is plain that without much consideration of the other factors affecting chemical reactions it would have been impossible to find a single set of numbers which would have fitted all cases. For instance, temperature and concentration would affect the course of a reaction as much as the nature of the constituents in many cases.

In the article *'Affinité'* in his *Encyclopédie méthodique* Guyton de Morveau reviewed critically all the attempts which had been made until the time of its publication (1787) to quantify the strength of affinity and chemical attraction, including his own attempt by the second method, which will be described below. He then proceeded to give numerical values for the affinities of various substances in much the same way as Fourcroy had done on the basis of their positions in the affinity tables, and to show these numbers in diagrams of double decompositions just as Fourcroy had done, so that the sums of the quiescent and divellent forces could be calculated.[31] However, he uses larger numbers than Fourcroy, so that fractions can be avoided, though he still gives no units and does not imply that the numbers have any physical meaning. For example, he represents the reaction between 'barytic muriate' (barium chloride) and 'mephite of potash' (potassium carbonate) thus:[32]

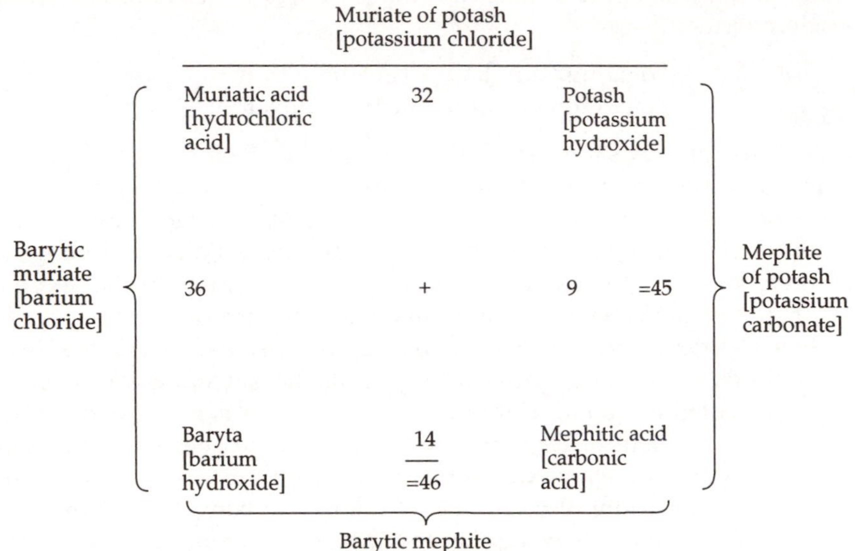

Fig. 5.5 Guyton de Morveau's illustration of the reaction between 'barytic muriate'(barium chloride) and 'mephite of potash'(potassium carbonate) to form 'muriate of potash'(potassium chloride) and 'barytic mephite'(barium carbonate), with his numeric values for the affinities involved. From L. B. Guyton de Morveau, *Encyclopédie méthodique: chymie*, Paris and Liège, 1786-9, p. 553.

The quiescent forces add up to 45 and the divellent forces to 46, so decomposition takes place and 'muriate of potash' and 'barytic mephite' (potassium chloride and barium carbonate) are formed. Guyton does not put the symbol for water in the middle of the diagram, as Bergman and Fourcroy did, to show that the reaction is in the wet way; but he shows by the shape of the horizontal brackets whether the products were left in solution, or as precipitates or sublimates, or in some other form. This is an extension of Bergman's device by which the point of the brackets was turned towards the centre of the diagram if the product remained in solution, or away from the centre if it was precipitated or otherwise removed from solution.

Attempts to measure the strength of affinity or chemical attraction by the second method, that is mechanically, may be traced back to Brook Taylor's measurements of the force of adhesion of a solid to a liquid, published in 1721.[33] These attempts are clearly founded on a theoretical assumption, though perhaps rather too simple a one, that the mechanical force which could be measured on the visible scale was directly related to the forces between insensibly small particles which produced chemical combination. In a paper read to the Royal Society in 1712, Taylor had described experiments like those of Hauksbee on the height to which liquids would rise when trapped between two glass plates, and from the heights compared the attractions of various liquids. In the paper of 1712 he went on to describe experiments in which he had hung thin pieces of firboard from one end of a pair of accurate scales and found what weights he had to place on the other end to lift the pieces 'at once' from the surface of stagnating water (that is, water which was smooth and undisturbed). He was encouraged to find that in every experiment the weight was exactly proportional to the surface—that is, presumably, to the surface of the board in contact with the water. However, he gave no evidence to show that the resistance of the board to being lifted was not due to atmospheric pressure.

In the very different climate of opinion of more than half a century later, in 1773, Guyton de Morveau carried out similar experiments. He showed that the adhesion of the solid to the surface of the liquid was not due solely to atmospheric pressure, for the force required to lift a glass disc from the surface of different liquids varied with the liquid, even though the atmospheric pressure was the same, and the glass disc adhered to the surface even under vacuum. A disc of tallow adhered to the surface of the water with various substances dissolved in it more strongly than glass, which was of particular interest because it had sometimes been suggested that there was a repulsion between fats and water.[34]

Assuming therefore that this method enabled him to measure the same force that was responsible for chemical combination, Guyton went on to examine the adhesion of discs of various metals, of a standard diameter of one inch, to the surface of mercury.[35] He found the following values for the weight necessary to lift each disc: gold 446 grains; silver 429 grains; tin 418 grains; lead 397 grains; bismuth 373 grains; zinc 204 grains; copper 142 grains; antimony 126 grains; iron 115 grains; cobalt 8 grains.

The fact that the order of the metals in this list was the same as the order of their affinity for mercury as shown in the appropriate column of an affinity table seemed to Guyton to show that the force which he had measured was that which was responsible for the solution of the metals in mercury. If so, he had at least for these particular cases got a measure of the force of chemical affinity itself. However, he did not proceed to calculate from the particular weights required to lift the discs in the particular circumstances of these experiments any absolute value for the forces involved.

In 1776 F. C. Achard (1753–1821), who though of French descent was Margraf's pupil and successor at the Berlin Academy, reported a similar series of measurements, though he used a wider range of liquids than Guyton.[36] Achard gave ten tables of the values for various substances, with corrections for differences of temperature, and deduced an equation for calculating the force of adhesion of an unknown substance of known specific gravity from its area of contact. Achard was a physicist as well as a chemist; but nevertheless this may be seen as a fairly early example of mathematics being applied to chemical concepts as opposed to physical concepts used to explain chemistry. However, he draws no chemical conclusions.

In his critical review of attempts to measure the force of affinity in his *Encyclopédie méthodique*, Guyton changed his opinion of the method of weighing the force of adhesion used by himself and Achard, since, being useless when the material of the solid disc dissolves readily in the liquid, it is applicable only in a few cases.[37] It was for that reason that he turned to the method of allocating numbers arbitrarily according to the position of substances in affinity tables which has already been described. In his account of other methods Guyton mentioned not only the work of Wenzel and Kirwan, which will be discussed below, but also the measurements by Lavoisier and Laplace of the heats absorbed in various chemical decompositions. With knowledge of the work on chemical thermodynamics which was to follow in the nineteenth century we may feel that Guyton showed brilliant foresight in connecting this work with variations in tendency to combine. However, his implication that Lavoisier and Laplace intended to give a measure of the affinities between the substances whose union was decomposed is misleading: they do not mention that possibility.[38]

The antecedents of the third method of measuring affinities, by inference from other measurable characteristics of chemical reactions, are in the work of W. Homberg (1652–1715), carried out in Paris, though he was of Saxon ancestry. Published in 1699 and 1700, these experiments are remarkable for their time in being thoroughly quantitative.[39] Like Lemery, Homberg quoted mechanical explanations in terms of the shapes of particles. In the paper published in 1699 Homberg described experiments to find the weights of actual acid in solutions of various specific gravities, so that the strength of an acid solution might be calculated from its specific gravity. This could be determined in a specific gravity bottle which he designed. He therefore determined the weights of spirit of nitre (nitric acid), spirit of salt (hydrochloric acid), oil of vitriol (sulphuric acid), *eau forte* [*sic*]—presumably more concentrated nitric acid—and distilled vinegar (acetic acid) which would combine with one ounce of salt of tartar (potassium carbonate). His method was to neutralize the acid with salt of tartar, evaporate the solution to dryness, and weigh the solid residue. Naturally he did not know that some weight had been lost through the escape of what would now be called carbon dioxide, and so he assumed that the difference in weight between the salt of tartar used and the solid residue was the weight of acid in the neutral salt.

Homberg then pointed out that since an ounce of good spirit of nitre contained twice as much acid as an ounce of good spirit of salt, there was no need to try to explain why *aqua regia* dissolved twice as much gold as spirit of salt by supposing that the points of the one were harder than the points of the other acid, and so separated the little particles of gold more strongly. Such Cartesian explanations were still in fashion.

In the paper published in 1700 he described how he had prepared samples of spirit of nitre and of spirit of salt (these being the two main types of acid) so that a given volume of each weighed as nineteen and seventeen respectively, compared with sixteen for an equal volume of river water. He then measured the weights of various earthy alkalis which one ounce of each acid spirit would dissolve promptly. The alkalis used were crabs' eyes, coral, pearls, mother of pearl, Oriental and Occidental bezoar, human calculus, oyster shells, calcined hartshorn, (all of which would be mostly calcium carbonate), quicklime (calcium oxide), and slaked lime (calcium hydroxide). Bole (clay, generally aluminium silicate mixed with various powdered rocks), *terra sigillata* (earthenware), and Tripoli earth (powdered quartz, mostly silicon oxide) were not dissolved by either acid. From the figures it was possible to calculate how much acid each kind of alkali retained, which would be valuable for physicians who had to prescribe these alkalis as medicines. Homberg found that quicklime and slaked lime had about the same ability to take up acids, so that their difference in causticity must be due to particles of fire which entered the quicklime when it was calcined.

Although the tendency of chemists to think in terms of precise weights grew during the century, nothing quite like this piece of work of Homberg's seems to have been done until the fourth quarter of the century. In a Newtonian paper which is largely concerned with the ways in which chemical attraction operated to produce neutral salts with properties different from the constituent acids and alkalis, originally read in 1739, Andrew Plummer had indeed reported quantitative experiments to establish the weights of various acids and alkalis in such salts. However, this is a rare instance, and Plummer does not go as far as Homberg in analysing the figures, nor suggest that the weights are a measure of the force of attraction.

In the 1770s such chemists as Cavendish and Lavoisier determined the actual weights of various substances which combined together. In particular Bergman tried to find the weights of phlogiston in various metals.[40] The mention of that in his *Dissertation on elective attractions*, together with his discussions of the influence of concentration on the extent to which a reaction in solution would go to completion, may have influenced C. F. Wenzel in his work on affinity. Wenzel believed that the order of affinity of substances for a solvent which would dissolve all of them was inversely as the time required for their solution. He constructed similar cylinders of various metals such as lead, gold, and silver, covered all but one face of each with melted amber, and allowed nitric acid to act on the uncovered face of each of them. From the weight of each which was dissolved in a given time he calculated the time that would be necessary for the whole of each cylinder to be dissolved. Wenzel gives no exact figures for the results of these experiments, though he does give figures for a number of analyses of salts elsewhere in the same book, and Professor Partington has even doubted whether the experiments described were ever carried out.[41]

Wenzel gives an analogy from mechanics. There the velocity which a body achieves in a given time is proportional to the force acting on it and inversely proportional to its mass. Wenzel therefore argued that the velocity of solution would depend on the nature of the metal, representing mass, and the affinity of the metal for the acid, representing force. Perhaps the reason why he can give only an order of affinities and not precise figures is that he cannot find a numerical measure of the nature of the metal, the analogue of mass, and so cannot use the times to calculate figures for the affinities.

Guyton de Morveau rejected Wenzel's assumption that the time of solution depended directly on affinity, since it would also depend on such factors as the cohesion of the particles of the metal (to which Wenzel might have replied that this would be included under the heading of 'the nature of the metal'), and because the order of the velocities was not always the same as the order of affinities.[42]

Objections to Wenzel's work were also made by the Irish chemist Richard Kirwan (1733–1812).[43] They were (1) that the alkalis and the earths were not dealt with by Wenzel (which was unavoidable, since his method was based on the solution of metals in acids); (2) that the results contradicted known facts, since although tin and regulus of antimony (metallic antimony) were quickly dissolved by the nitric acid, and lead and copper more slowly, it was well known that the affinity of nitric acid for copper was greater than it was for antimony; and (3) that other acids give different results. The last objection is not very persuasive, for it was normally assumed that in affinity tables the order of affinity for metals might be different for different acids.

Wenzel's notion that the speed of a reaction was governed by the affinity of the substances reacting recalls the remarks of Homberg; but there is no necessity to assume that he was influenced by them.

Kirwan based his own determination of the strength of affinity of acids for alkalis and bases on two principles. 'Firstly, that the quantity of real acid required to saturate a given weight of each basis is inversely as the affinity of each basis to such acid. Secondly, that the quantity of each basis requisite to saturate a given quantity of each acid is inversely as the affinity of such acid to each basis.'[44] In the first of his series of three papers on this subject (in which as well as referring to Wenzel he follows Homberg in some respects) he assumes that the weights of nitrous, vitriolic, and marine acids required to neutralize—or in eighteenth-century terms to saturate—a given quantity of fixed alkali are the same, as Homberg had supposed.[45] However, in the third paper, after Berthollet had criticized that assumption, Kirwan takes as the quantity of 'real acid' the weights in their potassium salts when evaporated to dryness. He then gives a table of the quantity of various bases which would saturate a hundred parts of vitriolic, nitrous, and marine (sulphuric, nitric, and hydrochloric) acids, which according to his assumptions would also give the relative strengths of the affinities of the bases for the acids.[46] Elliot reprinted this table in the second volume of his *Elements of the branches of natural philosophy connected with medicine* in 1786.

Kirwan also entered into the discussion of affinities in his *Essay on phlogiston* of 1784, in which among other arguments against the oxygen theory he suggested that the interpretations of several reactions required by the theory conflicted with the accepted order of affinities. As has already been mentioned, Professor Partington has connected Lavoisier's avoidance of a full discussion of affinities in the introduction to his *Traité élémentaire* with these difficulties, which Guyton admitted were embarrassing and demanded further attention.[47] However, William Higgins answered the objections in 1789.[48]

None of these attempts to measure affinity or chemical attraction quantitatively was generally accepted or even drew very much attention. The reason was no doubt largely that although each of the various sets of figures produced the right predictions for the results of some reactions, they led to the wrong predictions for others. That was inevitable if only one factor was taken into account. In fact the compilers of these sets of figures were in a dilemma. Either they allocated numbers without any theoretical reasons, and exposed themselves to the criticism that their tables were not founded on any rational basis, or else they derived their figures from experiments and some theoretical reasoning, and found that the results did not fit the facts. The former course might perhaps have on its side the argument that if figures which fitted the facts could be found they could lead to the perception of some general pattern, or some general law. However, not even by allocating numbers arbitrarily and readjusting them could a set of figures be found which fitted all reactions.

Another reason for the lack of general approval for these attempts, however, must have been that none of the methods suggested gave results in fundamental units or an absolute measure of the force of chemical attraction. Some of the results were in pure numbers without units, and others gave simply a measure of the relative affinity of several substances for another. Thus it was difficult to proceed to any theoretical conclusions. Even Guyton's method of weighing did not give a means for calculating the actual chemical forces involved, even if they were taken to represent chemical attraction rather than the effects of surface tension, and Wenzel's method gave still less information. Kirwan's figures represent combining weights, and suggest no method by which the forces that made substances combine could be calculated from them. It is obvious to a modern chemist using hindsight that Kirwan could have used his figures to work out a table of equivalents if he had recognized a pattern of regularity in them. In fact, however, the pattern is not at all obvious unless the experimenter is looking for it for other reasons, and it was left to Richter and Fischer to take that step.

Attempts at explanations through affinity or attraction

I shall now discuss some examples of the use made by eighteenth-century chemists of the concepts of chemical affinity or attraction to describe and explain their observations.

In Bergman's *Dissertation on elective attractions* the following is Scheme Number 20 in his series of diagrams of reactions involving more than one simple elective attraction, shown in words for the convenience of modern readers rather than in symbols as Bergman had it.[49]

Bergman's exposition of this diagram provides a good example of a description of a reaction in terms of the concept of chemical attraction:

> Scheme 20. Pl. I. exhibits the decomposition of calcareous hepar [i.e. a solution of calcium sulphide or polysulphides] by the vitriolic [sulphuric] acid. On the left side appears the hepar, indicated by the signs of its proximate principles united; but within the vertical bracket these principles are seen separate, one above the other. On the right, opposite the calcareous earth [i.e. lime] is placed the sign of vitriolic acid; in the middle stands the sign of water, indicating that the three surrounding bodies freely exercise their attracting powers in it. Now, as vitriolic acid attracts calcareous earth more forcibly than sulphur does, it destroys the composition of the hepar; the extruded sulphur being in itself insoluble falls to the bottom, which is signified by the point of the lower horizontal half-bracket's being turned downwards; and as the new compound, vitriolated calcareous earth (gypsum) [calcium sulphate] also subsides, unless the quantity of water be very large, the point of the upper bracket is likewise turned downwards.[50]

The sense of this explanation would not be altered if instead of the words 'attracts calcareous earth more forcibly' we read 'has a stronger tendency to combine with calcareous earth'; and the virtue of the exposition is that it gives a clear picture of the way in which the three radicals

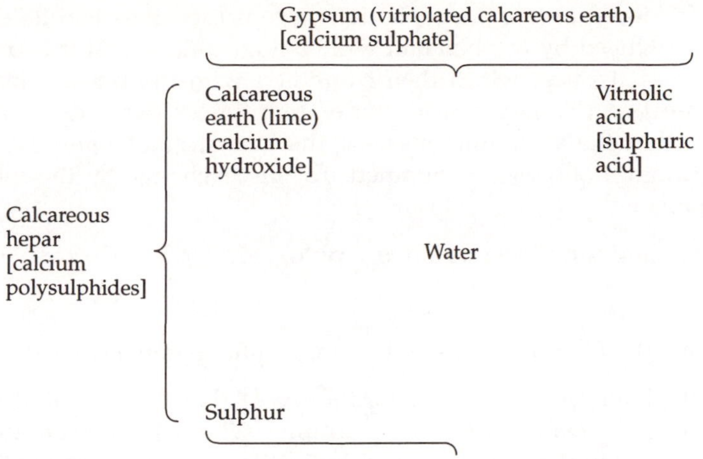

Gypsum (vitriolated calcareous earth)
[calcium sulphate]

Calcareous earth (lime) [calcium hydroxide]

Vitriolic acid [sulphuric acid]

Calcareous hepar [calcium polysulphides]

Water

Sulphur

Fig. 5.6 Bergman's Scheme No. 20 in his series of diagrams of reactions involving more than one simple elective attraction, expressed here in words rather than in symbols: the decomposition of calcareous hepar (a solution of calcium sulphide or polysulphide) by vitriolic (sulphuric) acid. From Tobern O. Bergman: *Dissertation on elective attractions*, London, 1970 (1785), Table I.

change their relationship but themselves remain unchanged, not that it offers any theoretical explanation of the cause of the change.

Lavoisier gives a similar explanation of the combination of phosphorus with oxygen gas, which he regarded as oxygen base in combination with caloric (his name for the matter of heat):

> This experiment proves, in the most convincing manner, that, at a certain degree of temperature, oxygen possesses a stronger elective attraction, or affinity, for phosphorus than for caloric; that, in consequence of this, the phosphorus attracts the base of the oxygen gas from the caloric, which, being set free, spreads itself over the surrounding bodies.[51]

Lavoisier's explanations of other cases of combustion are similar. As J. R. R. Christie has shown, as early as the 1750s Cullen had related thermal changes to chemical reactions and had attempted by measuring them to understand differences in chemical attraction. His view of the nature of the ether and its role in chemical combination was not unlike Lavoisier's view of caloric. However, Cullen's influence on Lavoisier was no doubt indirect, through the work of Joseph Black and perhaps Irvine and Crawford.

Such explanations were occasionally used by physiologists trying to give chemical explanations of biological processes. For instance, Adair Crawford (1749-95) suggested that animal heat was due to the fact that the attraction of the air for phlogiston was stronger than that of the blood for phlogiston, so that the phlogiston in the blood left it to combine with the air.[52] However, the atmospheric air from the lungs had its capacity for heat reduced by combination with phlogiston, so that it had to give up some of its heat, which then combined with the blood. This blood then absorbed phlogiston from the extremities of the body as it circulated, and at the same time released the heat it had acquired from the air. Partington and McKie summed up this exchange by the following equation:[53]

Atmospheric air + venous blood = phlogisticated air + arterial blood,

or,

(air + heat) + (blood + phlogiston) = (air + phlogiston) + (blood + heat).

In both Lavoisier's and Crawford's use of the concepts of affinity or attraction it is again evident that nothing would have been altered if 'tendency to combine' had been substituted, except of course that the words 'affinity' and 'attraction' show that the writer classes the effects which he is describing along with other examples of tendency to combine which were customarily referred to by those words and to which certain general characteristics were ascribed.

However, there were some phenomena which obliged eighteenth-century chemists, in searching for a theory to cover them, to go rather beyond the cautious empiricism which they usually preferred, and to picture a mechanism of what was happening in their reaction vessels. Some even went so far as to imagine mechanisms involving unobservably small particles; but speculation about the actual properties of the particles was kept to a minimum, except by a few. Although there is therefore a contrast between the bold speculative systems of attractions and repulsions by which some mathematicians and physicists tried to explain all phenomena, and the caution of most chemists, it is instructive to examine the explanations given by chemists when dealing with these topics. The phenomena which generally caused difficulty to chemists, though some non-chemists trying to give mechanical explanations for chemistry did not realize the importance of them, were the selectiveness of chemical affinity or attraction—that is, the fact that a substance would combine more readily with some substances than some others, and not at all with many others—and saturation (or, in later terms, neutralization in accordance with the law of constant composition)—that is, the fact that substances which did combine would do so only in certain proportions.

As we have seen, one or two explanations of the selectiveness of chemical affinity had been proposed which depended on the shapes or polarity of particles;[54] but none of them had been much noticed by chemists. They were able only to record as facts the 'orders of preference' of substances for each other. Rather more interest, however, was shown in attempts to explain saturation, and we shall therefore summarize briefly the history of the concept and some of the eighteenth-century attempts by chemists to account for saturation. This will involve the concept of chemical combination's being the result of an equilibrium between opposing forces.

Saturation

It must have been realized very early that a given amount of one substance would react only with a certain amount of another, for the idea is implicit in the quoting of definite quantities for the ingredients required in pharmaceutical preparations. The realization that measurement of quantities was important for chemical theory gained ground slowly during the eighteenth century, as we have seen, and several references to saturation occur.

Even in the seventeenth century Van Helmont mentions the saturation of alkali by a definite quantity of nitric acid (*quantum saturando alcali sufficit*—'as much as suffices for saturating the alkali').[55] Tachenius, who believed that everything in the universe was made up of acid and alkali, and interpreted their union in terms of love and hatred, seems to have been the first to refer to a definite point of saturation.[56] Boyle refuted the notion that everything consisted of acid and alkali; but he did take up the idea of a point of saturation. He used vegetable dyes as indicators in what were virtually titrations to show the point at which an acid exactly saturated an alkali.[57] He also wrote of menstruums being 'glutted' with what they dissolved.[58] Hooke similarly explained the inability of an enclosed portion of air to support combustion after a certain point as due to the glutting of air with the products of combustion.[59] Mayow also gave a good description of saturation.[60]

Thus by the end of the seventeenth century the idea of a definitely observable saturation-point, especially in the reaction between an acid and an alkali, was well established; but little attempt had been made to explain saturation. Such figurative words as 'saturated' or 'glutted', implying that chemical substances combined because they had appetites like living things, and ceased to combine because their appetites were satisfied, would tend to disguise the need for further explanations. However, the spread of the mechanical philosophy to chemistry caused the need for rational explanation to be felt.

Nicholas Lemery, for instance, discussed saturation in mechanical terms.[61]

Acid is always a solvent, when it is put in large enough quantity on the material which it is intended to dissolve; but it always produces a solid lump (*coagulum*) when being in too small a quantity its points are stuck in the pores of the material, and have not the force to separate them so as to get out. This is well seen when one pours some spirit of vitriol on the liquor of salt of tartar [i.e. sulphuric acid on potassium carbonate]; for if one puts on only what is necessary to penetrate the salt, the acid points stay as if trapped and make the substance heavy, which is why it brings about coagulation and precipitation. But if one adds to the liquor as much again or more of the spirit of vitriol as had been put on originally, the lump will disappear, because the little corpuscles which in combination were withstanding the acid and preventing its movement will be separated and dissolved by the acid, which has become stronger.

Newton, having introduced the theory that chemical action was due to attractive and repulsive forces between particles, mentions saturation briefly in the thirty-first Query at the end of the *Opticks*. He says that salt of tartar (potassium carbonate) does not draw more water out of the air than in a certain proportion to its quantity 'for want of an attractive force after it is satiated with Water'.[62] This suggests a more sophisticated model than Lemery's, but no explanation. However, a possible mechanical explanation was that a given particle of one substance would attract only enough particles of another to surround it, so that there was no room for more particles of the second substance to get close to it. It would then be saturated, and further particles of the second substance which approached it would not be able to get near enough to the particle of the first substance to be firmly attracted. Newton hinted at this possibility. He made it clear that chemical attraction fell off rapidly with distance. 'Since Metals dissolved in Acids attract but a small quantity of the Acid, their attractive Force can reach but to a small distance from them;[63] Others were later to take this hint. Knight and Boscovich also attempted mechanical explanations.[64]

Attention was drawn to saturation by Rouelle in the middle of the eighteenth century by his work on the three possible states of saturation of salts, i.e. acid, neutral, and basic salts.[65] The discussion was taken up by Macquer, who like most chemists of his time simply took it as an observed fact that the tendencies of substances to combine could be satisfied by a certain quantity of another substance, and would then cease to operate further.[66] He distinguished between relative saturation, where the tendency to combine is only diminished but not wholly satisfied, and absolute saturation, where it is wholly satisfied. This was in opposition to Rouelle, who supposed that there were definite points of saturation for each of the stages of combination between acids and alkalis that formed acid, neutral, and basic salts. Macquer also failed, as was normal at that time, to distinguish between the saturation of a solution, where no definite new compound is formed, and the saturation that marks the formation of a distinct new compound.

Macquer makes the interesting remark that 'the solution of bodies is not perfect, unless each of the integrant parts of one body is united to one of the integrant parts of another'.[67] In the article on Gravity (*Pesanteur*) he also offers an explanation for certain substances' being more easily satisfied than others, speculating in the manner of Buffon (in a volume published in the year before Macquer's *Dictionnaire*) about the way in which the laws of gravitation that apply to very large bodies may be modified when they are applied to very small particles in contact. Particles which are very dense, but which have only a small area of contact, will have only a small part of their tendency to unite satisfied, and substances composed of such particles will be very reactive. Such appear to be mineral acids, and in general all chemical solvents.

But if the parts of the dissolved body are of such a density, or have such close contact with the particles of the solvent, that the reactivity of the latter is entirely satisfied by their union, then the solvent will be in a state of rest, or in a kind of equilibrium, and will have no more dissolving power. This is the state or point of saturation. 'To make this saturation more complete, each of the integrant parts of the solvent must have met an integrant part of the dissolved body, upon which it may exhaust all its activity.'[68]

Both Boyle and Newton, and a little later Stahl, who no doubt directly influenced Macquer, conceived of a hierarchy of particles, the primary particles, which were the smallest, combining most strongly into small clusters, these clusters combining a little less strongly into slightly larger clusters, and so on.[69] However, the chemical principles or elements which were actually producible in the laboratory might be composed of such clusters, and would thus not be truly simple substances in the philosophical sense. It is this picture which Macquer has adapted to try to explain saturation. If his line of thought had been pursued further, and linked with some of the quantitative work on the combination of acids and metals which such chemists as Cavendish and Wenzel were doing, Dalton might have been anticipated by twenty or thirty years; but there were conceptual barriers. For a mid-eighteenth-century chemist it was dangerous to build very much on that sort of speculation. Dalton's method of weighing the unweighable, by inferring the relative weights of atoms from their combining weights, was undreamed of.

In the last quarter of the century, it is true, speculation was a little freer, especially in France. Chemists had achieved independence and were more confident. Guyton de Morveau, for instance, in his article on affinity in his *Encyclopédie méthodique*, quotes Buffon's discussion of the way in which the shape of particles might modify the effect of the law of gravitation, and Macquer's remarks on the same point.[70] Guyton then goes on to discuss how a single uniform law of attraction might explain the phenomena of cohesion, adhesion, and crystallization, and the variation of affinities from substance to substance, so that it would not be necessary to suppose that there were several different kinds of attraction. He then gives an example of how the effective forces of attraction between two solid tetrahedra, each composed of ten globules of equal size representing their 'primitive elements' and floating in a fluid of the same density, so that the gravity of the earth would not affect them, would vary according to the relative positions of the two tetrahedra. The forces of attraction are calculated numerically. However, Guyton does make it clear that this is merely an example of how the hypothesis put forward by Buffon might be developed in detail, and not a theory of how actual molecules are arranged. He introduces the calculation by the words:

But I do not wish to conclude this paragraph without trying to render still more readily appreciable the influence of shape in close attractions, at least to convey a conception of the extent to which it can produce varieties of effects which one would not have suspected of being capable of being engendered by the same force.[71]

Later on in the same article, Guyton discusses saturation without referring directly to this speculation. He follows Macquer rather than Rouelle.

Are there really different degrees of saturation of the same salt? Or rather is not the union which it contracts with the portion exceeding saturation only the effect of an additional combination (*surcomposition*) as with a foreign third substance? This is a question which deserves all our attention, not only because it concerns the general theory of chemical attraction, but further because of the necessity of knowing precisely the class of affinity involved before thinking of submitting it to calculation, or even of deducing from it a satisfactory explanation of the phenomena which depend on it.

I confess that the idea of different degrees of saturation of a substance by a single substance seems to me repugnant to all the notions which we have acquired up to the present of the manner in which combinations are formed. I readily conceive that the point of saturation of water by a salt can change according to its temperature, that colder water can absorb more acid gas; the increase or decrease of the matter of heat in the solvent changing the relative position, the density, perhaps the shape of the particles, it is not surprising that their attractive force is also modified by these changes; that they give rise to a more or less perfect contact; and that the force of affinity becomes capable, by this means, of removing from the influence of the law of gravitation a greater quantity of matter in one case than in the other; but we have nothing similar in the hypothesis under discussion; the circumstances are the same, the point of saturation cannot vary, since it is merely the effect of a cause which does not change.[72]

It is interesting to see how Guyton has used a theoretical model, which had been constructed to fit another set of phenomena, to interpret the phenomena of acid, neutral, and basic salts, and to make predictions about the nature of these new phenomena. His interpretation turned out not to fit the facts when they were better established; but his method of proceeding is admirable.

Guyton next states that chemists usually understand by saturation 'that state of a compound in which one of its component parts cannot accept or retain in combination a greater quantity of the other'.[73] According to that definition, it would be a contradiction to affirm that a substance could be saturated by two different quantities of another. Guyton, however, is not merely arguing about words. He goes on to show how reactions in which a salt appears to have two different states of saturation can be explained by the possibility that even when the acid and the base are combined in a saturated compound there may be a residual affinity between the compound itself and a further quantity of one of the components.

He illustrates this by the observation that when barium sulphate is boiled with sulphuric acid a small quantity of the salt dissolves, showing that it has some affinity with the acid. The reaction is further elucidated by showing it in a diagram, like Bergman's schemes of double decompositions, and expressing the affinities numerically. For this purpose Guyton does not use the numerical values of affinities which he suggests elsewhere in the same article, and which have been discussed above, since they refer to the main affinities, such as that between the base and the acid composing the salt. Instead he uses smaller numbers for the residual affinities, putting the affinity of the barium sulphate for the acid as equal to 4, and that of the acid for water as equal to 6. The reaction is then expressed as follows:[74]

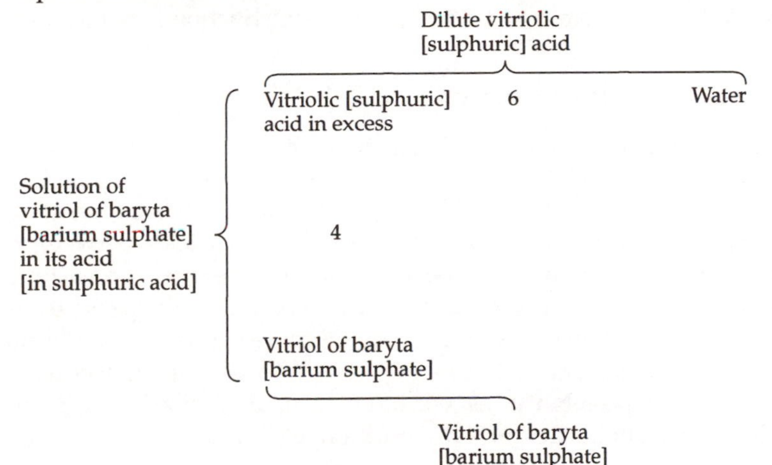

Fig. 5.7 Example of the affinity of a neutral salt with an excess of its acid, and of the separation of this excess by another simple affinity of composition. Guyton de Morveau's illustration of the effect of water in precipitating vitriol of baryta (barium sulphate) from its solution in vitriolic acid (sulphuric acid), showing his use of numeric values for residual affinities. From L. B. Guyton de Morveau, *Encyclopédie Méthodique: chymie*, Paris and Liège, 1786-9, p. 561.

The meaning of this is that water will precipitate the vitriol of baryta (barium sulphate) from its solution in vitriolic acid (sulphuric acid), because the affinity of 4 which combined the salt with the acid in solution is less than the affinity of 6 between the excess acid and the water.

This account is obviously developed from Bergman's ideas on the effects of excess of one of the reagents on a reaction. Guyton goes on to discuss five types of reaction in which an apparent second stage of saturation is to be explained by such factors.[75] The point for our present purpose is that Guyton does not resort to mechanical explanations of the kind which Buffon's ideas suggested, but confines himself in discussing saturation to purely chemical concepts and to what can be observed experimentally.

Even such speculation about the unobservable properties of particles as Macquer and Guyton allowed themselves was generally avoided, at least in print, by the more severe empiricists of the Scottish school. Saturation is noticed by Plummer[76] and by Black,[77] who distinguishes between simple saturation, where the limitation is on the quantity of only one of the ingredients (as when salt is dissolved in water), and reciprocal saturation, where neither of the ingredients can be combined with the other except in one proportion, as in the case of a neutral salt. Black also distinguishes between solution and true chemical reaction. In the same passage he remarks that a true chemical compound is distinguished by the mutual saturation of the components, whereas a mixture is possible with any proportion, though the distinction is not absolute.

The equilibrium of forces

In their efforts to explain saturation, chemists turned to the notion that it might be due to a state of equilibrium between two opposing forces or sets of forces. This could, of course, be done without making any assumptions about the shapes or sizes of the particles between which attraction or repulsion might be supposed to operate. The earliest suggestion that chemical combination is the result of the superiority of one set of opposing forces over another seems to have been Plummer's, though ultimately derived from Newton.[78] Cullen in his lectures mentioned such a possibility, as we have seen. A similar suggestion was made by Black in his doctoral dissertation. He wrote:

> The attraction of the volatile alkali for acids is stronger than that of *magnesia*, since it separated this powder from the acid to which it was joined. But it also appears, that a gentle heat is capable of overcoming this superiority of attraction, and of gradually elevating the alkali, while it leaves the less volatile acid with the *magnesia*.[79]

However, Black does not seem to be suggesting that the force of chemical attraction itself varies with heat. He also discusses the part played by competing attractions in a double decomposition. The precipitation of magnesia from a strong solution of magnesium chloride by quicklime (calcium oxide), he says, suggests that there is a stronger attraction between the calcareous earths (i.e. calcium oxide) and acids than between acids and magnesia. Yet if magnesia saturated with fixed air (i.e. magnesium carbonate) is mixed with a compound of acid and calcareous earth, the acid joins with the magnesia, and the calcareous earth with the fixed air (carbon dioxide).

Is it because the sum of the forces which tend to join the magnesia to the acid and the calcareous earth to the air, is greater than the sum of the forces which tend to join the calcareous earth to the acid and the magnesia to the air; and because there is a repulsion between the acid and the air, and between the two earths; or are they somehow kept asunder in such a manner as hinders any three of them from being united together?

The first part of this supposition is favoured by our experiments, which seem to show a greater difference between the forces wherewith the calcareous earth and magnesia attract fixed air, than between those which dispose them to unite with the acid. The repulsions hinted at in the second are perhaps more doubtful, though they are suggested in many other instances of decomposition.[80]

As we have seen, Cullen and Black used diagrams to illustrate such double decompositions.

The suggestion that chemical combination is the result of the superiority of one set of conflicting forces over another was also made by Macquer in the article *Affinité* in his *Dictionnaire de chymie*.[81]

Black's caution in the passage quoted above over the action of chemical repulsion is typical of his caution over anything that might appear speculative or hypothetical, a caution that increased later in his career. Plummer, to whom Black is probably referring, had actually suggested that chemical reaction depended on a conflict between chemical attraction and repulsion; but it was more common to attribute repulsion to the matter of heat, which might interfere with chemical effects, as did Lavoisier for instance. The work of Baumé on repulsion, and the elaborate use made of repulsion in the systems of Knight, Boscovich, and Higgins, have already been mentioned.

The possibility of an equilibrium between competing forces is considered by Black in his lectures, perhaps following Cullen, when he is discussing the crystallization of a salt from a super-saturated solution on moving the flask or on seeding.

A thermometer will show the emission of latent heat. There is in this experiment an equilibrium in the mixture, between its chemical attraction for latent heat, or the force with which it retains a certain quantity of heat in that form, and the cohesive attraction which tends to make it crystallize. These two attractions, the chemical and the cohesive, are always in opposition to one another, and here they are exactly balanced, or at least the force of the chemical attraction for heat, exceeds by very little the force of the cohesive which tends to crystallisation.[82]

As has been mentioned above in connection with Fourcroy's estimates of the force of chemical affinities, Kirwan invented the names 'quiescent' and 'divellent' for the forces opposing each other in a reaction.

In all these decompositions we must consider, first, the powers which resist any decomposition, and tend to keep the bodies in their present state; and secondly, the powers which tend to effect decomposition and a new union. The first I shall call quiescent affinities, and the second sort divellent. A decomposition will always take place when the sum of the divellent affinities is greater than that of the quiescent; and on the contrary, no decomposition will take place when the sum of the quiescent affinities is superior to, or equal to, that of the divellent: all we have to do, therefore, is to compare the sum of these powers.[83]

Bergman also discussed the conflict between competing attractions, and in an algebraic treatment which I have reviewed above (p. 138) considered the equilibrium which must exist between two opposed states at a particular temperature.[84]

In that passage Bergman came near to the Law of Mass Action, but was unable to go further because he had no way of testing his suggestion of an equilibrium by experiment. He was speaking only in terms of forces, which he could not measure, and had not made the further step of taking into account the concentration or relative mass of the reagents, which he might have measured. Also, he was not considering an equilibrium between three or four substances in solution, with competing forces of chemical combination reaching a balance: he was considering rather the competition between volatility and forces making for chemical combination.

Elsewhere, he does indeed discuss the possibility that the concentration of a substance might influence the result of a reaction. He clearly distinguishes between the proportion of a substance that is actually included in the resulting compound, which he assumes to be fixed, and the quantity of the substance which may have to be present in solution to make a reaction go to completion. In discussing the method by which he determined the order of affinities, he writes: 'It should be carefully noted in general, that there is occasion for twice, thrice, nay sometimes six times more of the quantity of the decomponent *c* than is necessary for saturating *A* when uncombined.'[85] On the other hand, when he does represent the result of a contest between the affinity of four substances for each other, producing a double decomposition, and shows it in a diagram, he deals only with the result of a completed reaction. Evidently there are conceptual barriers. The nearest he comes to considering an equilibrium in his diagrams of double decompositions is in numbers 9 to 11, in which he shows by little arrows on two sides of a substance that after the reaction it is partitioned between two other substances.[86]

One barrier is that he supposes chemical attraction to be at least analogous to gravitational attraction, and therefore proportional to the whole mass of the attracting substance present. If that were so, it would be natural to assume that the force making for combination as well as the effect of its concentration in making the reaction go to completion would be proportional to the mass of the substance present. In that case, the proportion of the substance in the resulting compound as well as the degree of completion of the reaction would be governed by concentration, and the only reason why one sample of a particular compound should have the same composition by weight as another would be that the solubility or some other physical property of one of its components tended to limit its concentration. That was precisely the argument of Claude Louis Berthollet (1748-1822), who in his *Recherches sur les lois de l'affinité* denied the law of definite proportions which had been tacitly assumed by earlier chemists.[87]

At the beginning of his book Berthollet criticized the belief of Bergman and others that 'elective affinity is an invariable force'.[88] He went on to discuss all the other factors, particularly concentration, that affected combining proportions as well as affinity. Nevertheless, he does in the course of his argument imply the assumption that fundamentally the affinity between substances is constant: it is merely its apparent effects which are altered by other factors.[89] A compound is formed when there is an equilibrium between the various forces and factors at work.[90] It is interesting that Berthollet allows himself in one place to picture the mechanism of chemical combination in terms of individual particles.[91]

Berthollet's views were controverted by Proust and others, but the line of thought followed by him, Bergman, Wenzel, and Richter led to eventual understanding of the Law of Equivalents and the Law of Mass Action. Bergman, who assumed in the orthodox way that compounds could exist in definite proportions, was on that account inhibited from following the argument as far as Berthollet. Also, without the concept of equivalent weights, Bergman could not properly compare the concentrations of different substances. However, both Bergman and Berthollet were inhibited, by their awareness of the lack of evidence about the behaviour of atoms, from writing publicly of chemical attraction or affinity as acting between one individual atom and another, except for the single passage in Berthollet referred to in the previous paragraph. They were therefore unable to distinguish between the effects of concentration on the velocity and degree of completeness of a reaction, and the reason for constant composition, that is the relative mass of the atoms concerned and their valency, in modern terms. Acceptable evidence to justify chemists in discussing the masses and arrangements of atoms was eventually produced by Dalton, but until then the speculations of such men as Bryan Higgins would have little influence.

Nevertheless Higgins, whose imaginative scheme is in strong contrast to the canny empiricism of such men as Black, Bergman, or even Macquer, was the first to give an intelligible explanation of saturation, apart from a few tentative hints. Indeed, Partington and Wheeler have traced a connection between the work of Bryan Higgins and his nephew William Higgins, who came near to anticipating Dalton.[92] Bryan Higgins pointed out that elements whose atoms attract atoms of their own kind could hardly saturate atoms of another kind which also attracted them and each other. He continued,

> Alkaline atoms repel each other; and when they stand distant from each other, cannot be approximated, without some power to counteract that of their repulsion. When they are drawn together by the attraction of water; as when we receive the elastic alkaline fluid [gaseous ammonia] in water, and thus compose the caustic volatile alkali [ammonia solution], the parts of water must be arranged alternately with parts of alkali; because each atom of water which attracts alkaline atoms, must be drawn into the line between them, whilst it draws them towards each other, contrary to their repulsive force.
>
> Of a determinate number of watery atoms, the sum of the attractive forces must be nearly equal to the repulsive forces of a certain number of alkaline atoms; and according to our principles, so much water, ought to condense or hold in combination so much alkali and no more. For if any additional portions of alkali be presented, they are indeed attracted by the water, but they are repelled by all the alkali which the water can hold; and as the parts of these portions repel each other, they must recede from each other, and from the water. And this inference from our principles, agrees with experience.
>
> But if it be true, that water is capable of being saturated with alkali, by reason of the mutual attraction of water and alkali, and of the reciprocal repulsion of alkaline atoms; and if our principles of attraction and repulsion be just, it is expected that, although water can be saturated by alkali, alkali cannot be saturated by water, so as to refuse union with any additional quantities of water. Because the parts of water, which attract each other as well as they attract those of alkali, can have no tendency to recede or seperate [*sic*] from the compounds of alkali and water. And we are not deceived in this expectation flowing from our premises ...[93]

Although his model of the mechanism of chemical combination is bolder and further removed from actual observation than Guyton's, Higgins uses it sensibly to make a prediction which can be compared with known phenomena. However, the weakness of his theory is that he can make no predictions that would suggest experiments to find phenomena still unknown by which the theory could be tested. Clearly he has been too ambitious in importing the mode of reasoning acceptable in speculative natural philosophy into chemistry.

The saturation between water and acid, water and air, earth and acid, earth and phlogiston, or any two elements of which one consists of atoms which repel each other according to Higgins's doctrine, is explained along the same lines. It is noticeable that like others of his time he does not make a total distinction between the saturation of a solution and the neutralization—as we now term it—which marks the formation of a definite compound. In the examples that he has given up to this point, the proportions are limited in only one direction—for instance, a given amount of alkali cannot all combine with less than a given amount of water, but it can combine with any greater amount of water.

The case of the mutual saturation of acid and alkaline matter is next considered and illustrated by a figure.[94] These two kinds of matter can combine only in certain definite proportions, which cannot be varied in either direction. Each sort of atom in such a case repels its own sort and attracts the other sort. The atoms will therefore be arranged alternately, so that each is surrounded by atoms of the other sort, and the attractive and repulsive powers exactly balance. Any additional acid or alkaline atom approaching the compound will find no uncompensated resultant attraction to draw it in, and so the compound can have only fixed proportions of its own constituents.

Such, then, was the use made of mechanistic theories by Bryan Higgins, which throws into relief the restraint of most other chemists in using such terms as 'attraction' or 'affinity' without venturing into the realms of hypothetical mechanisms. Even for those who did speculate, there were only two more or less tenable ways of explaining why the attraction of one body should be saturated by a certain quantity of another. One was the way hinted at by Macquer, for instance, of supposing that after a certain number of particles of the second substance had taken up all the space round each particle of the first, any additional particles of the second substance would be unable to ge near enough to be attracted by the first.[95] The other way is that taken by Higgins of supposing a balance to bet set up between hypothetical forces of attraction and repulsion. There was one pupil of Joseph Black's, William Keir (1753–83), who after reading Higgins actually suggested in his doctoral dissertation on chemical attraction that such attraction would be decreased as the quantity of the substance attracted increased, since the attractive power would have to be spread out over more of it. However, that can scarcely be classed as a tenable explanation.[96]

Envoi

The main theme of this book has been the gradual emancipation of chemists from the tutelage of the old-established discipline of physics, and particularly of the mechanical and corpuscular philosophy. In the course of that emancipation they rose to social and academic respectability and independence, partly because their work was seen as economically valuable, and partly perhaps because the most exciting discoveries in physics seemed to be in the past and the striking new developments seemed to be happening in chemistry. During this progress to autonomy chemists developed their own distinctive concepts, different from those of physics, in terms of which they tackled their own distinctive tasks. They believed firmly, as indeed most physicists claimed to believe, that theories should not be based on purely intellectual speculation but should be developed, if at all, by induction from observed phenomena. The phenomena should therefore be reduced to order, so that general patterns could be identified in them. Nevertheless, the notion that matter consisted of particles which though insensibly small had unchanging properties, such as size, shape, and mass, helped chemists to think of chemical reactions as involving specific and unchanging substances separating and combining. Hence they were able during the eighteenth century to form in effect their own list of chemically simple substances, quite different from the traditional elements or principles. This list was eventually made explicit and related to a new, empirical definition of an element and to a new, autonomous pattern of chemistry by Lavoisier.

Affinity tables were helpful in providing a rough classification of substances which had an empirical basis, and contributed to Lavoisier's eventual formulation of his list of simple substances. Yet much of the reason for their persistent popularity was the unfulfilled hope that they might indicate general patterns and lead to general laws from which the course of future chemical reactions might be predicted without recourse to theoretical speculation. Because they had no secure and reliable evidence that would have allowed them to accept any of the speculative theories about the causes and mechanisms of chemical combination that were constructed by physicists, chemists reduced the meaning of the words 'affinity' and 'chemical attraction' merely to 'tendency to combine'. Although this also helped them to form a mental model of the nature of chemical reactions through which to order the known phenomena and to find and accommodate more phenomena, it also in practice enabled them to avoid enquiring further into the philosophical causes of chemical reactions until long after the end of the eighteenth century.

Although they were able to classify types of affinity or attraction, or rather of circumstances in which affinity or attraction operated, they were not successful in attempts to measure affinity numerically or to infer laws that governed the operation of affinity or attraction.

It would be easy to conclude that affinity was a blind alley and a waste of time. That would be short-sighted. Eighteenth-century chemists thought that affinity or attraction was important; and even if the concept had eventually to be discarded, they used it a great deal in the course of thinking which did have successful results in other directions. To understand that thinking we have to understand what they meant and implied by affinity and chemical attraction. Certainly in studying its history we can attain insight into the processes by which chemistry evolved into an independent and valued branch of learning.

Notes

1. See R. E. Schofield, *Mechanism and materialism*, Princeton, 1970.
2. See p. 44 above.
3. R. Kirwan, *Philosophical Transactions of the Royal Society*, 1782, p. 179, and 1783, p. 15.
4. Macquer, *Elémens de chymie théorique*, 2nd edn, Paris, 1753 (1749), pp. 19-20.
5. Macquer, ibid., pp. 20–23.
6. Stahl, *Opusculum chymico-physico-medicum*, Halle, 1715, p. 232.
7. R. Boyle, *Works*, ed. Birch, London, 1744, Vol. II, p. 470; I. Newton, *Opticks*, New York, 1952 (1730), p. 394.
8. See R. Davy, *L'apothicaire Antoine Baumé (1728-1804)*, Cahors, 1955, pp. 38-40.
9. Baumé, *Manuel de chymie*, 2nd edn, Paris, 1765 (1763), pp. 7–8.
10. Baumé, ibid., pp. 8-14.
11. Macquer, *Dictionnaire de chymie*, Paris, 1766, Vol. I, pp. 49-54. See L. J. M. Coleby, *The chemical studies of P. J. Macquer*, London, 1938, esp. pp. 38-41.
12. Baumé, *Chymie expérimentale et raisonnée*, 3 vols, Paris, 1773, Vol. I, pp. 23–8.
13. E.g. Guyton de Morveau, *Elémens de chymie*, Dijon, 1777, pp. 79-97, simplified in *Encyclopédie méthodique*, Paris and Liège, 1786–9, Vol. I, pp. 548 ff.
14. Bergman, *Dissertation on elective attractions*, trans. Beddoes, London, 1970 (1783).
15. F. A. C. Gren, *Principles of modern chemistry* (trans. from *Grundriss der Chemie*, 2 vols, Halle, 1796–7), 2 vols, London, 1800, vol. I, pp. 47–9. The language and spelling of the eighteenth-century translator have been retained.
16. Gren, ibid., pp. 48-9.
17. Gren, ibid., p. 50.
18. Bergman, *Dissertation on elective attractions*, trans. Beddoes, London, 1970 (1783), pp. 18-29. See p. 138 above.
19. E.-F. Geoffroy, 'Table des différens rapports observés en chimie entre différentes substances', *Mémories de l'Académie Royale des Sciences*, 1718, pp. 202–12 (published 1719).
20. A. F. de Fourcroy, *Elémens d'histoire naturelle et de chimie*; first edn entitled *Leçons élémentaires d'histoire naturelle et de chimie*, Paris, 1782, 2 vols, Vol. I, pp. xxxiv-lxxii; 2nd edn, 4 vols, Paris, 1786; 3rd edn, 5 vols, Paris, 1789; 4th edn, 5 vols, Paris, 1791; 5th edn, 5 vols, Paris, 1793. Quoted with slightly altered wording by J. R.Partington, *History of chemistry*, Vol. IV, London, 1964, pp. 574–5.
21. See W. A. Smeaton, *Fourcroy, chemist and revolutionary, 1755–1809*, Cambridge, 1962, p. 97, referring to Lavoisier, *Mémoires de l'Académie Royale des Sciences*, 1777 (published 1780), pp. 592–600.
22. The change from the word *'affinité'* to the word *'attraction'* in the Laws was made in the 3rd edn of the *Elémens d'histoire naturelle et de chimie* of 1789 (pp. 43–72), although the word 'affinité' was still used occasionally as an alternative to 'attraction' elsewhere in that edition. The 3rd edn contains a discussion (pp. 62–6) of double elective attraction, clearly based on Bergman; refers to Kirwan's distinction between quiescent and divellent affinities; and

also refers to Fourcroy's own discussion in his *Mémoires et observations*. In this edn the second and third laws of the original set have been omitted, with consequent renumbering, although on p. 66 a reference to 'this tenth law' has accidentally been left in; but the two new laws have not yet been inserted. Nor do they occur in the 4th or 5th edns of the *Elémens*. The two new laws seem to occur for the first time in Fourcroy's *Systéme de connaissance chimiques*, 10 vols, Paris, 1800, Vol. I, pp. 67 ff; trans. by William Nicholson, *A general system of chemistry*, 11 vols, London, 1804, pp. 978-1021.

23. Guyton de Morveau, *Encyclopédie méthodique*, Paris and Liège, 1786–9, Vol. I, pp. 567–75.
24. J. Elliot, *Elements of the branches of natural philosophy connected with medicine*, London, 1782; 2nd ed, London, 1786, table headed 'Bergman's second table'.
25. Fourcroy, *Mémoires et observations de chimie*, Paris, 1784, pp. 308-23.
26. Fourcroy, ibid., p. 309.
27. Fourcroy, ibid., pp. 311–12.
28. Fourcroy, ibid., pp. 314–15 and 433–4.
29. Fourcroy, ibid., pp. 316, 318-21, and 323.
30. Fourcroy, ibid., pp. 430 (wrongly numbered 438) to 438.
31. Guyton de Morveau, *Encyclopédie méthodique*, Paris and Liège, 1786–9, Vol. I, pp. 552–9.
32. Guyton de Morveau, ibid., p. 553.
33. Brook Taylor, *Philosophical Transactions of the Royal Society*, xxvii, 1712, p. 538, and xxxi, 1721, pp. 204–8. Referred to by A. Thackray, 'Quantified chemistry—the Newtonian dream', in D. S. L. Cardwell (ed.), *John Dalton and the progress of science*, Manchester, 1968, pp. 96–7, and by W. A. Smeaton, 'Guyton de Morveau and chemical affinity', *Ambix*, **11**, 1963, pp. 55–64.
34. Guyton de Morveau, *Rozier's Observations sur la Physique*, **1**, 1773, pp. 172–7.
35. Guyton de Morveau, *Elémens de chymie*, Dijon, 1777, Vol. I, pp. 61 ff.
36. F. C. Achard, *Chymisch-physische Schriften*, Berlin, 1780, pp. 354–67.
37. Guyton de Morveau, *Encyclopédie méthodique*, Paris and Liège, 1786–9, pp. 576 ff.
38. Lavoisier and Laplace, 'Mémoire sur la chaleur', *Mémoires de l'Académie Royale des Sciences*, 1780, p. 355 (read 18 June 1783; published 1784).
39. G. Homberg, 'Observation sur la quantité exacte des sels volatiles acides contenus dans tous les différens esprits acides', and 'Observations sur la quantité d'acides absorbéz par les acalis terreux', *Mémoires de l'Académie Royale des Sciences*, 1699, pp. 44–51, and 1700, pp. 64–71.
40. Bergman, *Dissertatio chemica de diversa phlogisti quantitate in metallis*, Uppsala, 1782; *Dissertation on elective attractions*, London, 1970 (1785), pp. 268-70. See p. 144 above.
41. C. F. Wenzel, *Lehre von der Verwandtschaft der Körper*, Dresden, 1800 (1777), pp. 28-30, 85–97; Partington, *History of chemistry*, Vol. IV, London, 1964, p. 576.
42. Guyton de Morveau, *Encyclopédie méthodique*, Paris and Liège, 1786–9, pp. 577 ff.
43. R. Kirwan, *Philosophical Transactions of the Royal Society*, 1783, pp. 37–8.
44. Kirwan, ibid., p. 38.
45. Kirwan, 'Experiments and observations on the specific gravities and attractive powers of the mineral acids', *Philosophical Transactions of the Royal Society*, 1781, pp. 7–41. It will be noticed that to Kirwan 'affinities' and 'attractive powers' are synonymous.

46. Kirwan, 'Conclusion of the experiments and observations on the specific gravities and attractive powers of the mineral acids', *Philosophical Transactions of the Royal Society*, 1783, pp. 15–84.
47. See pp. 158–9 above and Chapter IV, Note 129.
48. W. Higgins, *A comparative view of the phlogistic and anti-phlogistic theories*, London, 1789.
49. Bergman, *Dissertation on elective attractions*, London, 1970 (1785), Table I.
50. Bergman, ibid., pp. 12–13.
51. A.-L. de Lavoisier, *Elements of chemistry*, trans. R. Kerr, Edinburgh, 1790, p. 57. See also J. R. R. Christie, 'Ether and the science of chemistry: 1740–1790', in G.N. Cantor and M. J. S. Hodge (eds), *Conceptions of ether: studies in the history of ether theories 1740–1900*, Cambridge, 1981, pp. 96–104.
52. A. Crawford, *Experiments and observations on animal heat*, London, 1779, pp. 68–73. Quoted by E. Mendelsohn, *Heat and life*, Cambridge, Mass., 1964, pp. 128–30.
53. J. R. Partington and D. McKie, 'Historical studies on the phlogiston theory', Part III, *Annals of Science*, **3**, 1938, p. 349.
54. For instance Newton (pp. 43–4 above), Boscovich (pp. 83–4 above), Lesage (p. 90 above), Demachy (pp. 90–91 above).
55. J. B. van Helmont, *Ortus medicinae*, Amsterdam, 1648, p. 56.
56. O. Tachenius, *Hippocrates chimicus*, Venice, 1666, p. 48.
57. Boyle, *Works*, ed. Birch, London, 1772, 2nd edn, Vol. II, pp. 53–4, 'Experiments ... touching colours', Experiments XX-XXII.
58. Boyle, ibid., Vol. III, p. 464.
59. R. Hooke, *Micrographia*, London, 1665, p. 104.
60. J. Mayow, *Medico-physical works* (Alembic Club reprint), Edinburgh and London, 1957 (1675), p. 167.
61. N. Lemery, *Cours de chymie*, 1756 (1675), pp. 733–4.
62. Newton, *Opticks*, New York, 1952 (1730), p. 377.
63. Newton, ibid., p. 395.
64. Knight, *An attempt to demonstrate that all the phaenomena in nature may be explained by ... attraction and repulsion*, 2nd edn, London 1754, pp. 19 ff; see p. 84 above. Boscovich, *Philosophiae naturalis theoria*, Vienna, 1758, pp. 37 ff; see p. 83 above. When atoms are at such a distance that they neither attract nor repel, saturation will result.
65. G. F. Rouelle, *Mémoires de l'Académie Royale des Sciences*, 1744, pp. 353–64.
66. Macquer, *Dictionnaire de chymie*, Paris, 1766, Vol. II, pp. 395–400.
67. Macquer, ibid., Vol. I, p. 353.
68. Macquer, ibid., Vol. II, p. 195.
69. Boyle, *The sceptical chymist*, London, 1661, p. 153; Newton, *Opticks*, New York, 1952 (1730), p. 394; Stahl, *Opusculum chymico-physico-medicum*, Halle, 1715, p. 232.
70. Guyton de Morveau, *Encyclopédie méthodique*, Paris and Liège, 1786–9, pp. 540–541.
71. Guyton de Morveau, ibid., pp. 546–7.
72. Guyton de Morveau, ibid., p. 560. Quoted by Seymour F. Mauskopf, 'Thomson before Dalton', *Annals of Science*, **25**, 1969, pp. 231–2.
73. Guyton de Morveau, *loc. cit.*
74. Guyton de Morveau, ibid., p. 561.
75. Guyton de Morveau, ibid., pp. 562–3.

76. A. Plummer, 'Experiments on neutral salts', *Essays and observations physical and literary*, 2 vols, 2nd edn, Edinburgh, 1771 (1756), Vol. I, pp. 372–3.
77. Black (ed. Robison), *Lectures on the elements of chemistry*, 2 vols, Edinburgh, 1803, p. 281.
78. A. Plummer, 'Experiments on neutral salts', *Essays and observations physical and literary*, 2 vols, 2nd edn, Edinburgh, 1771, (1756), Vol. I, pp. 374–5.
79. Black, *Experiments upon magnesia alba*, Alembic Club Reprint, Edinburgh, 1944 (1756), p. 44.
80. Black, *loc. cit.*
81. Macquer, *Dictionnaire de chymie*, Paris, 1766, Vol. I, pp. 49–54.
82. Black (ed. Robison), *Lectures on the elements of chemistry*, 2 vols, Edinburgh, 1803, Vol. I, p. 362.
83. Kirwan, *Philosophical Transactions of the Royal Society*, 1783, p. 40.
84. Bergman, *Dissertation on elective attractions*, London, 1970 (1785), pp. 14–16.
85. Bergman, ibid., p. 65.
86. Bergman, ibid., Table I.
87. C. L. Berthollet, *Recherches sur les lois de l'affinité*, Paris, 1801; *Researches into the laws of chemical affinity*, trans. M. Farrell, Baltimore, 1809 (1804), repr. 1966. See M. P. Crosland, *The Society of Arcueil: a view of French Science at the time of Napoleon I*, Cambridge, Mass., 1967.
88. Berthollet, *Researches into the laws of chemical affinity*, Baltimore, 1809, p. 4.
89. Cf. Berthollet, ibid., pp. 34–5, 75, 109–10, for instance.
90. Berthollet, ibid., p. 74.
91. Berthollet, ibid., p. 16.
92. J. R. Partington and T. S. Wheeler, *The life and work of William Higgins, Chemist (1763–1825)*, London, 1960. See Chapter 3, pp. 87–89 above.
93. B. Higgins, *A philosophical essay concerning light*, London, 1776, pp. 195–6.
94. Higgins, ibid., pp. 201–5 and Figure I at the end of the book. Quoted by Partington, 'Origins of the Atomic Theory', *Annals of Science*, **4**, 1939, pp. 269–72.
95. Macquer, *Dictionnaire de chymie*, Paris, 1766, Vol. II, p. 195.
96. William Keir, *De attractione chemica*, Edinburgh, 1778. See A. M. Duncan, 'William Keir's *De attractione chemica* (1778) and the concepts of chemical saturation, attraction and repulsion', *Annals of Science*, **23**, 1967, pp. 149–73.

Bibliography

Sources

F. C. Achard, *Chymisch-physische Schriften*, Berlin, 1780.

M. Alberti, *Physices propositiones et positiones fundamentales*, Halle, 1721.

Albertus Magnus, *Liber mineralium*. [No place or date, but probably printed by Cornelis de Zinkzea at Cologne about 1499.]

Anonymous, *Nouveau cours de chymie suivant les principes de Newton et de Sthall*, 2 vols, Paris, 1723; 2nd edn; Paris 1737.

Anonymous, *A narrative of the life and death of John Elliott, M.D., containing an account of the rise, progress, and catastrophe of his unhappy passion for Miss Mary Boydell*, London, 1787.

F. Bacon, Lord Verulam, *Novum organon*, London, 1620.

F. Bacon, Lord Verulam, *New Atlantis*, London, 1624.

A. Baumé, *Manuel de chymie*, Paris, 1763; 2nd edn, Paris, 1765.

A. Baumé, 'Réflexions sur l'attraction et la répulsion qui se manifestent dans la cristallisation des sels', *Avant-Coureur*, 16 March, 1772, p. 728, and *Rozier's Observations sur la physique*, **1**, 1773, p. 1.; repr. in R. Davy, *L'apothicaire Antoine Baumé*, Cahors, 1955, pp. 121–2.

A. Baumé, *Chymie expérimentale et raisonnée*, 3 vols, Paris, 1773.

A. Baumé, 'Expériences sur la décomposition du tartre vitriolé par l'acide nitreux seule', *Mémoires ... presentées à l'Académie Royale des Sciences par divers Sçavans*, **6**, 1774, pp. 231–6.

J. J. Becher, *Physica subterranea*, Leipzig, 1667; 4th edn, Leipzig, 1738.

Jean Béguin, *Tyrocinium chymicum*. [Anonymous. No place or date, but probably Paris, 1610.]

Jean Béguin, *Tyrocinium chymicum*, Cologne, 1611; a repr. of 1610 edn, also without the author's name.

Jean Béguin, *Tyrocinium chymicum*, Cologne, 1614.

Jean Béguin, *Les elémens de chymie*, Paris, 1615 (first, augmented, French edn of *Tyrocinium chymicum*.)

Jean Béguin, *Tyrocinium chymicum* (with the additional material from the 1615 French edn *Les élémens de chymie* trans. into Latin by Jeremy Barth). [No place, no date, but probably Frankfurt-an-der Oder and certainly 1618.]

Jean Béguin, *Tyrocinium Chymicum*, ed. Christopher Gluckradt; Königsberg, 1618; imprint: *Apud Iohannem Fabricium*'.

Jean Béguin, *Tyrocinium chymicum*, Königsberg, 1618. (Similar to the previous edn, but with imprint *Impensis Clementis Bergeri*.)

Jean Béguin, *Les élémens de chymie*, Rouen; [n.d., but probably 1624].

Jean Béguin, *Tyrocinium chymicum*, ed. Pelshoefer, Wittenberg, 1634. The Pelshoefer edn.

Jean Béguin, *Tyrocinium chymicum*, ed. Pelshoefer, Venice, 1643.

Jean Béguin, *Tyrocinium chymicum*, ed. Pelshoefer, Venice, 1669 (Similar to the 1643 edn.)

Jean Béguin, *Tyrocinium chymicum*, ed. Pelshoefer [incorporating some of his notes], Wittenberg, 1650.

Jean Béguin, *Tyrocinium chymicum*, Wittenberg, 1656; repr. of 1650 edn.

Jean Béguin, *Les élémens de chymie*, Lyons, 1658.

Jean Béguin, *Les élémens de Chymie*, Lyons, MDCXLXVI [*sic*]. (The date should probably be 1656, in which case this is the earlier Lyons impression.)

Jean Béguin, *Tyrocinium chymicum*, ed. Gerardus Blasius, Amsterdam, 1659.

Jean Béguin, *Tyrocinium chymicum*, ed. Gerardus Blasius, Amsterdam, 1669. (Another version of the 1659 edn.)

Jean Béguin, *Tyrocinium chymicum or Chymicall essayes* (English transl), London, 1669.

(Copies of many other editions survive.)

Torbern O. Bergman, Disquisitio de attractionibus electivis, *Nova Acta Regiae Societatis Scientiarum Upsaliensis*, **2**, 1775, pp. 161–250.

Torbern O. Bergman, *Dissertation on elective attractions*, trans. of 1775 version by J. A. Schufle, New York and London, 1968.

Tobern O. Bergman, *Opuscula physica et chemica*, 3 vols, Uppsala, 1779–83.

Torbern O. Bergman, *Dissertation on elective attractions*, (English trans., probably by Thomas Beddoes, of the version of 'Disquisitio de attractionibus electivis' in Vol. III of *Opuscula physica et chemica*, 1783), London and Edinburgh 1785; repr. with introduction and appendices by A. M. Duncan, London, 1970.

Torbern O. Bergman, Physical and chemical essays, (English trans. Edmund Cullen of most of *Opuscula physica et chemica*), 3 vols: I and II, London, 1784 and 1788; III, Edinburgh, 1791.

Torbern O. Bergman, *Dissertatio chemica de diversa phlogisti quantitate in metallis*, Uppsala, 1782.

Torbern O. Bergman, 'Meditationes de systemate fossilium naturali', *Nova Acta Regiae Societatis Upsaliensis*, **4**, 1784, pp. 63–128.

Torbern O. Bergman, *Foreign correspondence*, ed. Göte Carlid and Johann Nordstrom, Stockholm, 1965.

George Berkeley, *Treatise concerning the principles of human knowledge*, Dublin, 1710.

John Berkenhout, *First lines of the theory and practice of philosophical chemistry*, London, 1788.

Jakob Bernoulli, *Werke*, Band I, Basle, 1969.

C. L. Berthollet, *Recherches sur les lois de l'affinité*, Paris, 1801.

C. L. Berthollet, *Essai de statique chimique*, 2 vols, Paris, 1803.

C. L. Berthollet, *Researches into the laws of chemical affinity*, trans. M. Farrell, Baltimore, 1809 (1804); repr. 1966.

Joseph Black, 'Experiments upon magnesia alba, quicklime, and some other alcaline substances', in *Essays and observations, physical and literary*, 2 vols, Edinburgh, 1756, Vol. II, pp. 157–225; repr. as *Alembic Club Reprint* No. 1, Edinburgh, 1944.

Joseph Black, *Lectures on the elements of chemistry*, ed. John Robison, 2 vols, Edinburgh, 1803.

Joseph Black, *Notes from Dr Black's Lectures on chemistry, 1767/8*, ed. D. McKie, published by ICI Ltd, 1966.

François de le Boë [Sylvius], *Opera medica*, Amsterdam, 1679.

Hermann Boerhaave, *A new method of chemistry*; trans. by Peter Shaw of the pirated edn of Boerhaave's lectures, 2 vols, London, 1727.

Hermann Boerhaave, *Elementa chemiae*, 2 vols, Leiden, 1732.

Hermann Boerhaave, *Elements of chemistry*; trans. by Timothy Dallowe of the authorized edn of the *Elementa chemiae*, 2 vols, London, 1735.
Hermann Boerhaave, *New method of chemistry*; trans. by Peter Shaw of the authorized edn of the *Elementa chemiae*, 2 vols, London, 1741.
Roger Boscovich, *Philosophiae naturalis theoria reducta ad unicam legem virium in natura existentium*, Vienna, 1758; Venice, 1763.
Roger Boscovich, *A theory of natural philosophy* (English trans. by J. M. Child, of *Philosophiae naturalis theoria reducta ad unicam legem virium in natura existentium*), 2nd edn, Cambridge, Mass. 1966.
Robert Boyle, *The sceptical chymist*, 1661, Everyman edn, London, 1944.
Robert Boyle, *Electricity and magnetism*, repr. for the British Association of 'Experiments and notes about ... electricity', London, 1675, and 'Experiments and notes about ... magnetism', London, 1676; Oxford, 1927.
Robert Boyle, *Works*, abridged by Peter Shaw, 3 vols, London, 1725.
Robert Boyle, *Works*, ed. Thomas Birch, 2nd edn, 6 vols, London, 1772 (1st edn, London, 1744).
Jean-Louis Leclerc, Comte de Buffon, *Histoire naturelle*, 34 vols, Paris, 1749–89.
Cadet le Jeune, *Instituts de chymie*, Paris, 1770; French trans. of J. R. Spielmann, *Institutiones chemiae*, Strasburg, 1763; 2nd edn, Strasburg 1766.
J. F. Cartheuser, *Elementa chemiae medicae dogmatico-experimentalis*, Halle, 1736.
J. F. Cartheuser, *Vermischte Schriften aus der Naturwissenchaft, Chymie und Arzeneigelehrheit*, Leipzig and Magdeburg: parts 1–4, 1756; part 5, 1757; part 6, 1758; complete, 1762.
H. Cavendish, 'Three papers containing experiments on factitious air', *Philosophical Transactions of the Royal Society*, 1766, pp. 141–83.
H. Cavendish, *Scientific papers*, 2 vols, London, 1921.
A. C. Clairaut, *Mémoires de l'Académie Royale des Sciences*, 1745, pp. 329–64, 529–48, 577–9, 583–7.
Etienne Bonnot de Condillac, *Oeuvres philosophiques*, ed. Georges Le Roy, 3 vols, Paris, 1947–51.
M. J. A. N. de Caritat, Marquis de Condorcet, *Oeuvres*, 12 vols, Paris, 1847–9.
Roger Cotes, *Hydrostatical and pncumatical lectures*, London, 1738.
Adair Crawford, *Experiments and observations on animal heat*, London, 1779.
A. F. Cronstedt, *Essay towards a system of mineralogy*, trans. Engeström, London, 1770 (1758).
W. Cullen, *Lectures on chemistry*, [manuscripts belonging to Dr W. A. Smeaton, to the library of the Royal College of Physicians of Edinburgh, and to the Wellcome Historical Medical Library].
J. F. Demachy, *Recueil de dissertations*, Amsterdam, 1774.
J. F. Demachy, *Précis d'une nouvelle table des combinations ou rapports*, Paris, (n.d., but *c.* 1770).
J. T. Desaguliers, *A course of experimental philosophy*, 2 vols, London, 1734.
R. Descartes, *Discours de la méthode*, Leiden, 1637.
R. Descartes, *Meditationes de prima philosophia*, Amsterdam, 1641.
R. Descartes, *Principia philosophiae*, Amsterdam, 1644.
D. Diderot and J. le R. D'Alembert, *Encyclopédie*, 35 vols, Paris, 1751–80.
R. Dossie, *The elaboratory laid open, or, The secrets of modern chemistry and pharmacy revealed*, London, 1758.
H. L. Duhamel du Monceau, 'Sur la base du sel marin', *Mémoires de l'Académie Royale des Sciences*, 1736, pp. 215–32.

H. Duhamel du Monceau and Grosse, 'Sur les différentes manïeres de rendre le tartre soluble', II, *Mémoires de l'Académie Royale des Sciences*, 1733, pp. 260–72.

John Elliot, *Elements of the branches of natural philosophy connected with medicine*, London, 1782; 2nd edn, London, 1786.

John Elliot, 'Observations on the affinities of substances in spirit of wine', *Philosophical Transactions of the Royal Society*, **76**, 1786, p. 155.

J. C. P. Erxleben, *Anfangsgründe der Chemie*, Göttingen, 1775.

[Fizes], Leçons de chymie de l'Université de Montpellier, Paris, 1750.

A. F. de Fourcroy, *Leçons élémentaires d'histoire naturelle et de chimie*, 2 vols, Paris, 1782 [first edn title of *Elémens d'histoire naturelle et de chimie*].

A. F. de Fourcroy, *Mémoires et observations de chimie*, Paris, 1784.

A. F. de Fourcroy, *Elémens d'histoire naturelle et de chimie*, 4th edn, 5 vols, Paris, 1791; 5th edn, 5 vols, Paris, 1793.

A. F. de Fourcroy, [Articles in] Vol. iii of *Encyclopédie méthodique*, 13 vols, Paris, 1786–1830.

A. F. de Fourcroy, 'Recherches sur le métal des cloches, et sur les moyens ...', *Annales de Chimie*, **9**, 1791, pp. 305–52.

A. F. de Fourcroy, *A general system of chemical knowledge*, (English trans. by William Nicholson of *Système de connoissances chimiques*, 10 vols, Paris, 1800), 11 vols, London, 1804.

A. F. de Fourcroy, 'Observations sur le tableau du produit des affinités chymiques', *Rozier's Observations sur la Physique*, 1773, **2**, pp. 197–204.

John Freind, *Chymical lectures in which almost all the operations of chymistry are reduced to their true principles and the laws of nature*, read in the Museum at Oxford, 1704; 2nd edn, London, 1729 (1712).

Galileo Galilei, *Opere*, 15 vols, Florence, 1842ff.

Galileo Galilei, *Tractatio de demonstratione*, ed. W. A. Wallace, Padua, 1988.

Galileo Galilei, *Tractatio de praecognitionibus et praecognitis*, ed. W. A. Wallace, Padua, 1988.

Geber, *Works*, trans. R. Russell, ed. E. J. Holmyard, London, 1928.

C. E. Gellert, *Anfangsgründe zur metallurgischen Chymie*, 2nd edn, Leipzig, 1776 (1751).

C. E. Gellert, *Metallurgic chemistry*, trans. I. S. (probably John Seifert), London, 1776.

C. F. Geoffroy, 'Observations sur la terre d'alun; manière de la convertir en vitriol, ce qui fait une exception à la Table des rapports en chymie', *Mémoires de l'Académie Royale des Sciences*, 1744, pp. 69–76.

E.-F. Geoffroy, 'Table des différens rapports observés en Chimie entre différentes substances', *Mémoires de l'Académie Royale des Sciences*, 1718, pp. 202–12, (published 1719).

E. -F. Geoffroy, 'Eclaircissement sur la table inserée dans les Mémoires de 1718', *Mémoires de l'Academie Royale des Sciences*, 1720, pp. 20–3.

E. -F. Geoffroy, *Treatise of the fossil, vegetable and animal substances that are made use of in physick*, trans. G. Douglas, London, 1736.

P. Gergens and S. Hochheimer, *Tabellen über die chemische Verwandtschaft der Körper*, Frankfurt-am-Main, 1790.

J. R. Glauber, Works, trans. Packe, London, 1688.

J. F. Gmelin, *Geschichte der chemie*, 3 vols, Göttingen, 1797–9.

T. Graham, *Elements of chemistry*, London, 1842.

W. J. S. van 'sGravesande, *Mathematical elements of natural philosophy*, trans. J. T. Desaguliers, London, 1720.

F. A. C. Gren, *Systematisches Handbuch der gesammten Chemie*, 2 vols, Halle, 1787–90; 2nd edn, 4 vols, Halle, 1794–6.

F. A. C. Gren, *Grundriss der Chemie*, (abridged version of *Systematisches Handbuch*), 2 vols, Halle, 1796–7.

F. A. C. Gren, *Principles of modern chemistry*, (trans. by Gruber of *Grundriss der Chemie*), 2 vols, London, 1800.

Grosse, 'Recherche sur le plomb', *Mémoires de l'Académie Royale des Sciences*, 1733, pp. 313–28.

Grosse and Duhamel, 'Sur les différentes manières de rendre le tartre soluble', II, *Mémoires de l'Académie Royale des Sciences*, 1733, pp. 260–72.

L. B. Guyton de Morveau, *Digressions académiques*, Dijon, 1772.

L. B. Guyton de Morveau, *Rozier's Observations sur la Physique*, 1773, **1**, pp. 172–7.

L. B. Guyton de Morveau, [Articles on] 'Affinité', 'Crystallisation', 'Dissolution', and 'Equipondérance' in Diderot's *Supplément à l'Encyclopédie*, 4 vols, Amsterdam, 1776–7.

L. B. Guyton de Morveau, *Elémens de chymie, théorique et pratique*, 3 vols, Dijon, 1777–8.

L. B. Guyton de Morveau, *Encyclopédie méthodique: chymie*, Paris and Liège, 1786–9.

John Hadley, *A plan of a course of chemical lectures*, Cambridge, 1758.

Stephen Hales, *Vegetable staticks*, London, 1727; Oldbourne repr., London, 1961.

John Harris, *Lexicon technicum*, 2 vols, London, 1704 and 1710.

David Hartley, *Observations on man*, London, 1749.

J. B. van Helmont, *Ortus medicinae*, Amsterdam, 1648.

J. F. Henkel, *Mediorum chymicorum non ultimum, conjunctionis primum, appropriatio*, Dresden and Leipzig, 1727.

J. F. Henkel, *Pyritologia*, (English trans.), London, 1757.

Bryan Higgins, *A philosophical essay concerning light*, Vol. I (only one volume issued), London, 1776.

William Higgins, *A comparative view of the phlogistic and anti-phlogistic theories*, London, 1789.

Thomas Hobbes, *Leviathan*, London, 1651.

F. Hoffmann, *Chemia rationalis et experimentalis, sive collegium physico-chymicum curiosum*. Leiden, 1748.

P.-H. T. d'Holbach, *Système de la nature*, Amsterdam, 1770.

G. Homberg, 'Observation sur la quantité exacte des sels volatiles acides contenus dans tous les différens esprits acides', *Mémoires de l'Académie Royale des Sciences*, 1699, pp. 44–51.

G. Homberg, 'Observations sur la quantité d'acides absorbéz par les alcalis terreux', *Mémoires de l'Académie Royale des Sciences*, 1700, pp. 64–71.

R. Hooke, *Micrographia*, London, 1665.

G. L. Hume, *Chemical attraction*, Cambridge, 1835.

I. Kant, *Gesammelte Schriften*, 22 vols, Berlin, 1902–42.

John Keill, *An introduction to natural philosophy or lectures read in the University of Oxford, Anno Dom. 1700*, London, 1720, 4th edn, London, 1745.

John Keill, *Philosophical Transactions of the Royal Society*, 1708, pp. 97–110.

John Keill, 'De operatione chymicarum ratione mechanica', trans. Anna Guerrini and Jole R. Shackelford, *Ambix*, **36**, 1989, pp. 138–52.

[James Keir] *A treatise on the various kinds of permanently elastic fluids or gases*, London, 1779.

Richard Kirwan, *Philosophical Transactions of the Royal Society*, **1781**, pp. 7–41; **1782**, pp. 179–236; **1783**, pp. 15–84.

Richard Kirwan, *Essay on phlogiston*, London, 1784; 2nd edn, with trans. by William Nicholson of notes from the French edn of 1788, London, 1789.

Richard Kirwan, *Essai sur le phlogistique*, trans. Mme Lavoisier, with notes by Guyton, Lavoisier, Laplace, Monge, Berthollet, and Fourcroy, Paris, 1788.

Gowin Knight, *An attempt to demonstrate that all the phaenomena in nature may be explained by two simple active principles, attraction and repulsion ...*, London, 1748; 2nd edn, London, 1754.

H. Kopp, *Geschichte der Chemie*, 4 vols, Brunswick, 1843–7.

A.-L. de Lavoisier, 'Observations sur quelques circonstances de la crystallisation des sels', *Rozier's Observations sur la physique*, 1773, 1, p. 1C.

A.-L. de Lavoisier, 'Mémoire sur la Combustion en général', *Mémoires de l'Académie Royale des Sciences*, 1777, p. 592. (Read 1777; published 1780).

A.-L. de Lavoisier, 'Mémoire sur l'affinité du principe oxygine avec les différentes substances auxquelles il est susceptible de s'unir', *Mémoires de l'Académie Royale des Sciences*, 1782, p. 530. (Read 1783; published 1785).

A.-L. de Lavoisier, *Traité elémentaire de chimie*, 2 vols, Paris, 1789.

A.-L. de Lavoisier, *Elements of chemistry*, trans. R. Kerr, Edinburgh, 1790.

A.-L. de Lavoisier and P. S. Laplace, 'Mémoire sur la chaleur', *Mémoires de l'Académie Royale des Sciences*, 1780, p. 355. (Read 1783; published 1784).

N. Lemery, *Cours de chymie*, ed. Théodore Baron d'Hénouville, Paris, 1756 (1675).

G. L. Lesage, *Essai de chymie méchanique*, Geneva, 1762.

William Lewis, *The new dispensatory*, London, 1753.

William Lewis, *Commercium philosophico-technicum*, London, 1763.

J. P. de Limbourg, *Dissertation sur les affinités chymiques*, Liège, 1761.

John Locke, *Essay concerning human understanding*, London, 1690; abridged edn by A. S. Pringle-Pattison, Oxford, 1924.

T. Lucretius Carus, *De rerum natura*, ed. C. Bailey, Oxford, 1947.

P.-J. Macquer, *Elémens de chymie-Théorique*, 2nd edn, Paris, 1753 (1749).

P.-J. Macquer, *Dictionnaire de chymie*, 2 vols, Paris, 1766; 2nd edn, Paris, 1778.

P.-J. Macquer, *Dictionary of chemistry*, trans. James Keir, 2 vols, London, 1771; 2nd edn, London, 1777.

A. S. Margraf, *Opuscules chymiques*, trans. Formey and Demachy, 2 vols, Paris, 1762.

P. A. Marherr, *Dissertatio chemica de affinitate corporum*, Vienna, 1762.

J. Marzucchi, *Chymiae elementa*, Padua, 1751.

John Mayow, *Medico-physical works*, trans. of *Tractatus quinque medico-physici*, 1675; published as Alembic Club reprint No. 17, Edinburgh and London, 1957.

De la Méthérie, *Observations sur la physique*, **34**, 1789, pp. 43–5.

J. F. Meyer, *Chymische Versuche*, Hanover and Leipzig, 1764; French trans., 2 vols, Paris, 1765.

J. Michell, *Treatise of artificial magnets*, Cambridge, 1750.

A. G. Monnet, *Traité de la dissolution des Métaux*, Amsterdam and Paris, 1775.

P. van Musschenbroek, *Epitome elementorum physico-mathematicorum in usus academicos*, Leiden, 1726; English trans. J. Colson, 2 vols, London, 1744.

P. van Musschenbroek, *Cours de physique expérimentale et mathématique*, trans. Sigaud de la Fond, Paris, 1769.

Caspar Neumann, *Chemical works*, trans. Lewis, London, 1759.

Isaac Newton, *Mathematical principles of natural philosophy*, trans. A. Motte (1729); revised F. Cajori, Berkeley, 1934.

Isaac Newton, *Opticks*, Dover repr. of 1730 edn, London and New York, 1952.

Isaac Newton, *Papers and letters on natural philosophy*, ed. I. B. Cohen, Cambridge, 1958.

Isaac Newton, *Correspondence*, ed. H. W. Turnbull. 4 vols, Cambridge, 1959–67.

Isaac Newton, *Unpublished scientific papers*, ed. A. R. and M. B. Hall, Cambridge, 1962.

Isaac Newton, *Four letters from Sir Isaac Newton to Doctor Bentley*, London, 1756; repr. in Hall and Hall (eds), *Unpublished scientific papers of Isaac Newton, Cambridge, 1962.*

[Isaac Newton], 'Traité des reflexions, refractions et inflexions des couleurs de la lumierè, [presumed to be Isaac Newton's Opticks], London, 1704 in 4 (en anglois), [in] *Catalogus librorum ... Stephani-Francisci Geoffroy*, Paris, 1731, No. 1502.

William Nicholson, *The first principles of chemistry*, London, 1790; 3rd edn, London, 1796.

William Nicholson, *A general system of chemistry*, 11 vols, London, 1804.

E. A. Nicolai, *De affinitate corporum*, Jena, 1775.

J. Page, 'Receipts for preparing and compounding the principal medicines made use of by the late Mr Ward', *Medical Museum*, **1**, London, 1763, No. XXV, pp. 255–78.

Paracelsus, *Werke*, ed. Sudhoff, 14 vols, Munich and Berlin, 1922–33.

George Pearson, *A translation of the table of chemical nomenclature*, 2nd edn, London, 1799.

Henry Pemberton, *A view of Sir Isaac Newton's philosophy*, London, 1728.

Henry Pemberton, *A course of chemistry*, ed. James Wilson, London, 1771.

Thomas Percival, 'Facts and queries relative to attraction and repulsion', *Memoirs of the Literary and Philosophical Society of Manchester*, **2**, 1785, pp. 429–39.

Andrew Plummer, *Opera chemica*. [Manuscript notes on lectures delivered at Edinburgh *c*.1750, now in the Library of the Royal College of Physicians.]

Andrew Plummer, 'Experiments on neutral salts', *Essays and observations physical and literary*, 2 vols, 2nd edn, Edinburgh, 1771. (Read 1738 and 1739.)

J. H. Pott, *Chymische Untersuchungen*, Potsdam, 1746; 2nd edn, Berlin, 1757.

J. H. Pott, *Lithéognosie*, trans. d'Holbach, Paris, 1753.

J. H. Pott, *Dissertations chymiques*, trans. Demachy, 4 vols, Paris, 1759.

Joseph Priestley, *Experiments and observations on different kinds of air*, 2nd edn, 3 vols, London, 1775–7 (1774–7).

Joseph Priestley, *Disquisitions relating to matter and spirit*, London, 1777.

J. Quincy, *Pharmacopée universelle raisonnée*, trans. Clausier, Paris, 1749.

W. Richardson, *The chemical principles of the metallic arts*, Birmingham, 1790.

Bryan Robinson, *A treatise of the animal oeconomy*, Dublin, 1732.

John Robison, *A system of mechanical philosophy*, ed. David Brewster, 4 vols, Edinburgh, 1822.

G. Rothen, *A synopsis, or short analytical view of chemistry*, trans. Alexander Macbean, London, 1743 (1717).

G. F. Rouelle, 'Mémoire sur les sels neutres', *Mémoires de l'Académie Royale des Sciences*, 1744, pp. 353–64.

D. A. Rüdiger, *Systematische Anleitung zur reinen und überhaupt applicirten oder allgemeinen Chymie*, Leipzig, 1756.

Sage, *Mémoires de chimie*, Paris, 1773.
C. W. Scheele, *Chemical essays*, trans. T. Beddoes, London, 1786.
C. W. Scheele, *Collected papers*, trans. L. Dobbin, London, 1831.
Ephraim Seehl, *A new improvement in the art of making the true volatile spirit of sulphur*, London, 1744.
[Senac], *Nouveau cours de chmie suivant les principes de Newton & Sthall*, 2 vols, Paris, 1723; 2nd edn, Paris, 1737.
P. Shaw, *Chemical lectures*, London, n.d.
J. R. Spielmann, *Institutiones chemiae*, Strasburg, 1763; 2nd edn, Strasburg, 1766; French trans. by Cadet le Jeune, *Instituts de chymie*, 2 vols, Paris, 1770.
G. E. Stahl, *Zymotechnia fundamentalis, seu Fermentationis theoria generalis*, Halle, 1697.
G. E. Stahl, *Opusculum chymico-physico-medicum*, Halle, 1715.
G. E. Stahl, *Fundamenta chymiae, Nuremberg, 1723.*
G. E. Stahl, *Ausführliche Betrachtung ... von den Saltzen*, Halle, 1723.
G. E. Stahl, *Philosophical principles of universal chemistry*, trans. of *Fundamenta chymiae* by Peter Shaw, London, 1730.
G. E. Stahl, *Traité du soufre*, trans. d'Holbach, Paris, 1766.
G. E. Stahl, *Traité des sels* (trans. by d'Holbach of *Ausführliche Betrachtung ... von den Saltzen*), Paris, 1771.
L. D. Suckow, *Entwurf einer phẏsischen Scheidenkunst*, Frankfurt and Leipzig, 1769.
E. Swedenborg, 'Principia rerum naturalium', in *Opera philosophica et mineralia*, Vol. I; 3 vols, Leipzig, 1734.
O. Tachenius, *Hippocrates chimicus*, Venice, 1666.
O. Tachenius, *Hippocrates chimicus* and *Clavis*, trans. J. W., London, 1677.
Brook Taylor, *Philosophical Transactions of the Royal Society*, **27**, 1712, p. 538, and **31**, 1721, pp. 204–8.
L. Tessari, *Chymiae elementa in aphorismos digesta*, Venice, 1772.
Thomas Thomson, *The history of chemistry*, 2 vols, London, 1830–1.
J. B. Trommsdorff, *Darstellung der Saüren, Alkalien, Erden und Metalle*, Erfurt, 1800.
Venel, Article 'Chymie' in the *Encyclopédie*, Vol. III, Paris, 1753; repr. Fourcroy in *Encyclopédie Méthodique*, Vol. III, Paris, 1796, p. 302.
François-Marie Arouet de Voltaire, *Les éléments de la philosophie de Newton*, Paris, 1738.
François-Marie Arouet de Voltaire, *The elements of Sir Isaac Newton's philosophy*, London, 1738.
Martin Wall, 'Some observations on the phaenomena, which take place between oil and water', and 'Extracts of two letters from Dr Wall of Oxford, to Dr Percival, in reply to the foregoing queries concerning attraction and repulsion', *Memoirs of the Literary and Philosophical Society of Manchester*, **2**, 1785, pp. 419–28, and 439–50.
John Warltire, *Analysis of a course of lectures in experimental philosophy*, 6th edn, London, 1769; with appendix 'Tables of the various combinations and specific attractions'.
D. C. E. Weigel, *Grundriss der reinen und angewandten Chemie*, 2 vols, Greifswald, 1777.
Carl Friedrich Wenzel, *Lehre von der Verwandtschaft der Körper*, Dresden, 1800 (1777).
J. C. Wiegleb, *Handbuch der allgemeinen Chemie*, 2 vols, Berlin and Stettin, 1781.
J. C. Wiegleb, *General system of chemistry*, trans. C. R. Hopson, London, 1789.

Secondary works

E. J. Aiton, *The vortex theory of planetary motions*, London, 1972.
E. J. Aiton, *Leibnitz: a biography*, Bristol, 1985.
A. Anastasi, *Nicolas Leblanc, Sa Vie, Ses travaux, et l'histoire de la soude artificielle*, Paris, 1884.
R. G. W. Anderson (ed.), *Joseph Black 1728–99*, Edinburgh, 1982.
R. G. W. Anderson and C. Lawrence (eds) *Science, medicine and dissent: Joseph Priestley (1733–1804)*, London, 1987.
R. G. W. Anderson and A. D. C. Simpson (eds) *The early years of the Edinburgh Medical School*, Edinburgh, 1976.
G. Averley, 'The "social chemist": English chemical societies in the eighteenth and early nineteenth century', *Ambix*, **33**, 1986, pp. 99–128.
John J. Beer, 'Eighteenth-century theories on the process of dyeing', *Isis*, **51**, 1960, pp. 21–30.
B. Bensaude-Vincent, 'A view of the chemical revolution through contemporary textbooks: Lavoisier, Fourcroy and Chaptal', *British Journal for the History of Science*, **23**, 1990, pp. 435–60.
M. Beretta, *The enlightenment of matter: the definition of chemistry from Agricola to Lavoisier*. Canton, Mass., 1993.
A. J. Berry, *Henry Cavendish*, London, 1960.
Marie Boas, 'Boyle as a theoretical scientist', *Isis*, 41, 1950, pp. 263–64.
Marie Boas, 'The establishment of the mechanical philosophy', *Osiris*, **10**, 1952, pp. 412–541.
Marie Boas, *Robert Boyle and seventeenth-century chemistry*, Cambridge, 1958.
Marie Boas, 'Newton's chemical papers', in *Isaac Newton's papers and letters on natural philosophy*, ed. I. B. Cohen, Cambridge, 1958, pp. 241–8. (*See also* Marie Boas Hall.)
G. Bowles, 'John Harris and the powers of matter', *Ambix*, **22**, 1975, pp. 21–38.
D. Brewster, *Memoirs of ... Sir Isaac Newton*, London and New York, 1965; repr. of Edinburgh 1855 edn, 2 vols.
P. W. Bridgman, *The logic of modern physics*, New York, 1927.
Pierre Brunet, *L'introduction des théories de Newton en France XVIIIe siècle*, Paris, 1931.
Gerd Buchdahl, *The image of Newton and Locke in the Age of Reason*, London, 1961.
D. S. L. Cardwell (ed.) *John Dalton and the progress of science*, Manchester, 1968.
H. Cassebaum, *Carl Wilhelm Scheele*, Leipzig, 1982.
J. R. Christie, 'The origins and development of the Scottish scientific community, 1680–1760', *History of Science*, **12**, 1974, pp. 122–41.
J. R. Christie, 'The rise and fall of Scottish science', in M. P. Crosland (ed.), *The emergence of science in Western Europe*, London, 1975, pp. 111–26.
J. R. R. Christie, 'Ether and the science of chemistry: 1740–1790', in G. N. Cantor and M. J. S. Hodge (eds), *Conceptions of ether: studies in the history of ether theories 1740–1900*, Cambridge, 1981, pp. 85–110.
Marshall Clagett (ed.), *Critical problems in the history of science*, Madison, 1959.
A. Clericuzio, 'A redefinition of Boyle's chemistry and corpuscular philosophy', *Annals of Science*, **47**, 1990, pp. 561–89.
A. Clow, 'Chemistry at the older universities of Britain during the 18th century', *Nature*, **155**, 1945, pp. 158–62.
A. Clow, 'Hermann Boerhaave and Scottish chemistry', in Andrew Kent (ed.), *An eighteenth-century Lectureship in Chemistry*, Glasgow, 1950, pp. 41–8.

A. and N. L. Clow, 'Vitriol in the Industrial Revolution', *Economic History Review*, **15**, 1945, pp. 44–55.

A. and N. L. Clow, *The chemical revolution*, London, 1952.

I. B. Cohen, *Franklin and Newton*, Philadelphia, 1956.

I. B. Cohen, 'Isaac Newton, Hans Sloane, and the Académie Royale des Sciences', in *Mélanges Alexandre Koyré*, 2 vols, Paris, 1964; Vol. I, pp. 61–116.

L. J. M. Coleby, 'Studies in the chemical work of Stahl', unpublished University of London Ph. D. thesis, 1928.

L. J. M. Coleby, *The chemical studies of P.J. Macquer*, London, 1938.

Jean-Paul Contant, *L'enseignement de la chimie au Jardin Royal des Plantes de Paris*, Cahors, 1952.

M. P. Crosland, 'The use of diagrams as chemical "Equations"', *Annals of Science*, **15**, 1959, pp. 75–90.

M. P. Crosland, *Historical studies in the language of chemistry*, London, 1962.

M. P. Crosland, *The Society of Arcueil: a view of French science at the time of Napoleon I*, Cambridge, Mass., 1967.

M. P. Crosland, 'Lavoisier's theory of acidity', *Isis*, **64**, 1973, pp. 306–25.

M. P. Crosland (ed.), *The emergence of science in Western Europe*, London, 1975.

M. P. Crosland, 'The image of science as a threat: Burke versus Priestley and the "Philosophic Revolution"', *British Journal for the History of Science*, **20**, 1987, pp. 277–307.

R. Davy, *L'apothicaire Antoine Baumé (1728–1804)*, Cahors, 1955.

A. G. Debus, *The chemical dream of The Renaissance*, Cambridge, 1968.

A. G. Debus (ed.), *Science, medicine and society in the Renaissance*, 2 vols, London, 1972.

H. Diels, *Die Fragmente der Vorsokratiker*, 4th edn, 2 vols, Berlin, 1922.

B. J. T. Dobbs, *The foundations of Newton's alchemy: the hunting of the Greene Lyon*, Cambridge, 1975.

B. J. T. Dobbs, 'Newton's alchemy and his theory of matter', *Isis*, **73**, 1982, pp. 511–28.

A. Donovan, *Philosophical chemistry in the Scottish Enlightenment*, Edinburgh, 1975.

A. Donovan, 'The chemical revolution: essays in reinterpretation', *Osiris*, 2nd series, vol. **4**, 1988.

P. Duhem, *Le mixte et la combinaison*, Paris, 1902.

J. B. Dumas, 'Rapport relatif à la découverte de la soude artificielle', *Compte rendu des séances de l'Académie des Sciences*, **42**, 1856, pp. 553–78.

A. M. Duncan, 'Some theoretical aspects of eighteenth-century tables of affinity', *Annals of Science*, **18**, 1962, pp. 177–94 and 217–32.

A. M. Duncan, 'William Keir's *De attractione chemica* (1778) and the concepts of chemical saturation, attraction and repulsion', *Annals of Science*, **23**, 1967, pp. 149–73.

A. M. Duncan, 'The functions of affinity tables and Lavoisier's list of elements', *Ambix*, **17**, 1970, pp. 28–42.

A. M. Duncan, 'Styles of language and modes of chemical thought', *Ambix*, **28**, 1981, pp. 83–107.

A. M. Duncan, 'Particles and eighteenth-century concepts of chemical combination', *British Journal for the History of Science*, **21**, 1988, pp. 447–53.

Jacob Ellowitz, 'The history of the theories of chemical affinity from Boyle to Berzelius', unpublished University of London M.Sc. dissertation, 1927.

Eduard Farber, 'Variants of preformation theory in the history of chemistry', *Isis*, **54**, 1963, pp. 443–60.

J. Fauvel, R. Flood, M. Shortland, and R. Wilson (eds), *Let Newton be!*, Oxford, 1989.

Karen Figala, 'Newton as alchemist', *History of Science*, **15**, 1977, pp. 102–37.

D. Geoghegan, 'Some indications of Newton's attitude towards alchemy', *Ambix*, **6**, 1957, pp. 102–6.

C. C. Gillispie, 'The discovery of the Leblanc process', *Isis*, **48**, 1957, pp. 152–70.

C. C. Gillispie, 'The *Encyclopédie* and the Jacobin philosophy of science: a study in ideas and consequences', in Marshall Clagett (ed.), *Critical problems in the history of science* (commentaries by H. B. Hill and H. Guerlac in the same volume), Madison, 1959, pp. 255–89.

J. V. Golinski, 'Peter Shaw: chemistry and communication in Augustan England', *Ambix*, **30**, 1983, pp. 19–29.

J. V. Golinski, 'Robert Boyle: scepticism and authority in seventeenth-century chemical discourse', in A. E. Benjamin, G. N. Cantor and J. R. R. Christie (eds), *The figural and the literal: problems of language in the history of science and philosophy, 1630–1800*, Manchester, 1987, pp. 58–82.

J. V. Golinski, 'Utility and audience in eighteenth-century chemistry: case studies of William Cullen and Joseph Priestley', *British Journal for the History of Science*, **21**, 1988, pp. 1–31.

J. V. Golinski, *Science as public culture: chemistry and enlightenment in Britain*, 1760–1820, Cambridge, 1992.

J. Graham, 'Revolutionary philosopher: the political ideas of Joseph Priestley (1733–1804)', *Enlightenment and Dissent*, **8**, 1989, pp. 43–68.

H. E. Le Grand, 'Lavoisier's oxygen theory of acidity', *Annals of Science*, **29**, 1972, pp. 1–18.

R. Grave, *The Scottish philosophy of common sense*, Oxford, 1960.

F. Greenaway, *John Dalton and the atom*, London, 1966.

Henry Guerlac, 'Commentary on the papers by C. C. Gillispie and L. Pearce Williams', in Marshall Clagett (ed.), *Critical problems in the history of science*, Madison, 1959, pp. 317–20.

Henry Guerlac, 'Newton in France—two minor episodes', *Isis*, **53**, 1962, pp. 219–21.

Henry Guerlac, 'Newton's optical aether', *Notes and Records of the Royal Society of London*, **22**, 1967, pp. 45–67.

Henry Guerlac, 'The background to Dalton's atomic theory', in D. S. L. Cardwell (ed.), *John Dalton and the progress of science*, Manchester, 1968, pp. 57–91.

Henry Guerlac, *A. L. Lavoisier: Chemist and revolutionary*, New York, 1975.

A. R. Hall, *The Scientific Revolution*, London, 1954.

A. R. Hall, *Philosophers at war: the quarrel between Newton and Leibniz*, Cambridge, 1980.

A. R. Hall, *The revolution in science 1500–1750*, (rev. edn of *The Scientific Revolution*, London, 1983.

A. Rupert Hall, *Isaac Newton: adventurer in thought*, Oxford, 1989.

Marie Boas Hall, 'The history of the concept of element', in D. S. L. Cardwell (ed.), *John Dalton and the progress of science*, Manchester, 1968, pp. 21–39.

M. Boas Hall, *British Journal for the History of Science*, **10**, 1977, pp. 262–4.

A. R. and M. B. Hall, 'Newton's chemical experiments', *Archives Internationales d'Histoire des Sciences*, **11**, 1958, pp. 113–52.

A. R. and M. B. Hall, 'Newton's theory of matter', *Isis*, **51**, 1960, pp. 131–44.

A. R. and M. B. Hall (eds), *Unpublished scientific papers of Isaac Newton*, Cambridge, 1962, pp. 321–31.

Owen Hannaway, 'Johann Conrad Barchusen (1666–1723)', *Ambix*, **14**, 1967, pp. 96–111.

Owen Hannaway, *The chemists and the word: the didactic origins of chemistry*, Baltimore, 1975.

Joan L. Hawes, 'Newton's revival of the aether hypothesis and the explanation of gravitational attraction', *Notes and Records of the Royal Society of London*, **23**, 1968, pp. 200–12.

P. M. Heimann, 'Nature is a perpetual worker', *Ambix*, **20**, 1973, pp. 1–25.

P. M. Heimann and J. E. McGuire, 'Newtonian forces and Lockean powers: concepts of matter in eighteenth-century thought', *Historical Studies in the Physical Sciences*, **3**, 1971, pp. 233–06.

H. B. Hill, 'Commentary on the papers by C.C. Gillispie and L. Pearce Williams', in Marshall Clagett (ed.), *Critical problems in the history of science*, Madison, 1959, pp. 309–16.

Ho Ping-Yü and J. Needham, 'Theory of categories in early medieval Chinese alchemy', *Journal of the Warburg and Courtauld Institutes*, **22**, 3–4, 1959, pp. 173–210.

F. L. Holmes, *Lavoisier and the chemistry of life: an exploration of scientific creativity*, Madison, 1985.

F. L. Holmes, *Eighteenth-century chemistry as an investigative enterprise*, Berkeley, 1989.

K. Hufbauer, *The formation of the German chemical community* (1720–1795), Berkeley and Los Angeles, 1982.

Lynn Sumida Jay, *Gassendi the atomist: advocate of history in an age of science*, Cambridge, 1988.

Milton Kerker, 'Hermann Boerhaave and the development of pneumatic chemistry', *Isis*, **46**, 1955, pp. 36–49.

D. Knight, *Ideas in chemistry: a history of the science*, London, 1992.

J. G. Knight, The chemical studies of Hermann Boerhaave, unpublished University of London M.Sc. dissertation, 1933.

A. Koyré, *Newtonian studies*, London, 1965.

T. S. Kuhn, 'Newton's "31st Query" and the degradation of gold', *Isis*, **42**, 1951, pp. 296–8.

T. S. Kuhn, 'Reply to Marie Boas', *Isis*, **43**, 1952, pp. 123–4.

T. S. Kuhn, 'The independence of density and pore-size in Newton's theory of matter', *Isis*, **43**, 1952, p. 123.

D. S. Landes, *The unbound Prometheus*, Cambridge, 1969.

T. H. Levere, *Affinity and matter: elements of chemical philosophy (1800–65)*, Oxford, 1971.

T. H. Levere, 'Thomas Beddoes at Oxford: radical politics and the Regius Chair in Chemistry', *Ambix*, **28**, 1981, p. 61–9.

T. H. Levere, 'Dr Thomas Beddoes (1750–1808): science and Radical Medicine in politics and society', *British Journal for the History of Science*, **17**, 1984, pp. 187–204.

J. A. L. Lemay and R. E. Oesper, 'The lectures of Guillaume François Rouelle', *Journal of Chemical Education*, **31**, 1954, pp. 338–43.

G. Lindeboom, *Hermann Boerhaave: the man and his work*, London, 1968.

D. J. Lysaght, 'Hooke's theory of combustion', *Ambix*, **1**, 1937, p. 103.

Russell McCormmach, 'Henry Cavendish: a study of rational empiricism in eighteenth-century natural philosophy', *Isis*, **60**, 1969–70, pp. 293–306.
J. G. McEvoy, 'Joseph Priestley, "Aerial Philosopher": metaphysics and methodology in Priestley's chemical thought, from 1762 to 1781', *Ambix*, **25**, 1978, pp. 1–55, 93–116, 153–75, and **26**, 1979, 16–38.
J. E. McGuire, 'Transmutation and immutability: Newton's doctrine of physical qualities', *Ambix*, **14**, 1967, pp. 69–95.
J. E. McGuire, 'Force, active principles and Newton's invisible realm', *Ambix*, **15**, 1968, pp. 154–208.
J. E. McGuire, 'The origin of Newton's doctrine of essential qualities', *Centaurus*, **12**, 1968, pp. 233–60.
J. E. McGuire, 'Atoms and the "analogy of nature": Newton's third rule of philosophizing', *Studies in History and Philosophy of Science*, **1**, 1970, pp. 3–58.
J. E. McGuire, 'Existence, actuality and necessity: Newton on space and time', *Annals of Science*, **35**, 1978, pp. 463–508.
J. E. McGuire and P. M. Rattansi, 'Newton and the "Pipes of Pan"', *Notes and Records of the Royal Society of London*, **21**, 1966, pp. 108–43.
D. McKie, 'Some notes on Newton's chemical philosophy', *Philosophical Magazine*, **33**, 1942, pp. 847–70.
D. McKie, 'John Harris and his *Lexicon Technicum*', *Endeavour*, **4**, 1945, pp. 53–7.
D. McKie, *Antoine Lavoisier*, London, 1952.
F. E. Manuel, *The religion of Isaac Newton*, Oxford, 1974.
P. Mathias, 'Who unbound Prometheus? science and technical change 1600–1800', *Yorkshire Bulletin*, **21**, 1969, pp. 3–16.
S. M. Mauskopf, 'Thomson before Dalton', *Annals of Science*, **25**, 1969, pp. 229–42.
C. Meinel, 'Theory or practice? The eighteenth-century debate on the scientific status of chemistry', *Ambix*, **30**, 1983, 121–32.
E. M. Melhado, 'Oxygen, phlogiston and caloric: the case of Guyton', *Historical Studies in the Physical Sciences*, **13**, 1983, pp. 311–34.
E. Mendelsohn, *Heat and life*, Cambridge, Mass., 1964.
Hélène Metzger, *Newton, Stahl, Boerhaave et la doctrine chimique*, Paris, 1930.
Hélène Metzger, *Attraction universelle et religion naturelle chez quelques commentateurs anglais de Newton*, Paris, 1938.
F. C. Millington, 'Studies in cohesion', *Lychnos*, **45**, 1944, pp. 57–78.
L. T. More, *Isaac Newton: a biography*, New York, 1962, (repr. of 1st edn, New York, 1934).
J. B. Morrell, 'The University of Edinburgh in the late eighteenth century: its scientific eminence and academic structure', *Isis*, **62**, 1971, pp. 158–71.
M. M. Pattison Muir, *A history of chemical theories and laws*, New York, 1907.
R. P. Multhauf, *The origins of chemistry*, London, 1966.
A. E. Musson and Eric Robinson, *Science and technology in the Industrial Revolution*, Manchester, 1969.
A. E. Musson and Eric Robinson, 'Science and industry in the late 18th century', *Economic History Review*, **13**, 1960, pp. 222–44.
S. H. Nasr, *Science and civilization in Islam*, Cambridge, Mass., 1968.
E. W. J. Neave, 'Chemistry in Rozier's Journal', Parts VIII and IX, *Annals of Science*, **8**, 1952, p. 28.
J. Needham, *Science and civilization in China*, Vol. V, Cambridge, 1974 ff.
D. Oldroyd, 'An examination of G. E. Stahl's *Philosophical principles of universal chemistry*', *Ambix*, **20**, 1973, pp. 36–52.

Margaret J. Osler, 'Locke and the changing ideal of scientific knowledge', *Journal of the History of Ideas*, **31**, 1970, pp. 3–16.

J. R. Partington, 'Origins of the atomic theory', *Annals of Science*, **4**, 1939, pp. 269–72.

J. R. Partington, *History of chemistry*, 4 vols, [but only Vols. II-IV and pt. 1 of Vol. I published], London 1961–71.

J. R. Partington and D. McKie, 'Historical studies on the phlogiston theory', *Annals of Science*, **2**, 1937, pp. 361–404; **3**, 1938, pp. 1–58 and pp. 337 ff.; **4**, 1939, pp. 113–49.

J. R. Partington and T. S. Wheeler, *The life and work of William Higgins, Chemist (1763–1825)*, Oxford and London, 1960.

T. S. Patterson, 'Jean Béguin and his *Tyrocinium chymicum*', *Annals of Science*, **2**, 1937, pp. 243–98.

Margula R. Perl, 'Newton, Leibniz and Clarke', *Journal of the History of Ideas*, **30**, 1969, pp. 507–26.

C. E. Perrin, 'Lavoisier's table of the elements: a reappraisal', *Ambix*, **20**, 1973, pp. 95–105.

C. E. Perrin, 'A reluctant catalyst: Joseph Black and the Edinburgh reception of Lavoisier's chemistry', *Ambix*, **29**, 1982, pp. 141–76.

P. M. Rattansi, 'Newton's alchemical studies', in A. G. Debus (ed.), *Science, medicine and society in the Renaissance*, London, 1972, Vol. II, pp. 183–98.

P. M. Rattansi, 'Some evaluations of reason in sixteenth- and seventeenth-century natural philosophy', in M. Teich and R. Young (eds), *Changing perspectives in the history of science*, London, 1973, pp. 148–166.

L. Roberts, 'Setting the table: the disciplinary development of eighteenth-century chemistry as read through the changing structure of its tables', in P. Dear (ed.), *The literary structure of scientific argument: historical studies*, Philadelphia, 1991, pp. 99–132.

L. Rosenfeld, 'Newton's views on aether and gravitation', *Archive for the History of the Exact Sciences*, **6**, 1969, pp. 29–37.

Charles Ross, 'Studies in the chemical work of Claude Louis Berthollet', unpublished University of London M.Sc. dissertation, 1933.

Margaret Rowbottom, 'The chemical studies of Robert Boyle', unpublished University of London Ph.D. thesis, 1955.

C. A. Russell, *Science and social change*, 1700–1900, London, 1983.

R. E. Schofield, 'Joseph Priestley, the theory of Oxidation and the nature of matter, *Journal of the History of Ideas*, **25**, 1964, pp. 285–94.

R. E. Schofield, *A scientific autobiography of Joseph Priestley*, Cambridge, Mass., 1966.

R. E. Schofield, 'Joseph Priestley, natural philosopher', *Ambix*, **14**, 1967.

R. E. Schofield, *Mechanism and materialism: British natural philosophy in an age of reason*, Princeton, 1970.

E. L. Scott, 'The Macbridean doctrine of air', *Ambix*. **17**, 1970, pp. 43–7.

W. L. Scott, 'The impact of the French Revolution on English science', in *Mélanges Alexandre Koyré*, 2 vols, Paris, 1964, Vol. II, ed. F. Braudel, pp. 475–95.

S. Shapin and S. Schaffer, *Leviathan and the air-pump: Hobbes, Boyle, and the experimental life*, Princeton, 1985.

W. Shea (ed.), *Revolutions in science*, Canton, Mass., 1988.

R. Siegfried, 'Lavoisier's table of simple substances: its origin and interpretation', *Ambix*, **29**, 1982, pp. 29–48.

R. Siegfried, 'Lavoisier and the phlogistic connection', *Ambix*, **36**, 1989, pp. 31–40.

Robert Siegfried and Betty Jo Dobbs, 'Composition, a neglected aspect of the Chemical Revolution', *Annals of Science*, **24**, 1968, pp. 275–93.

Nathan Sivin, 'William Lewis (1708–1781) as a chemist', *Chymia*, **8**, 1962, p. 75.

W. A. Smeaton, 'Macquer on the composition of metals and the artificial production of gold and silver', *Chymia*, **11**, 1966, pp. 81–8.

W. A. Smeaton, 'L. B. Guyton de Morveau (1737–1816) his life and works', unpublished University of London M.Sc. dissertation, 1953.

W. A. Smeaton, 'The contributions of P. J. Macquer, T. O. Bergman, and L. B. Guyton de Morveau to the reform of chemical nomenclature', *Annals of Science*, **10**, 1954, pp. 87–106.

W. A. Smeaton, *Fourcroy: chemist and revolutionary, 1755–1809*, London, 1962.

W. A. Smeaton, 'Guyton de Morveau and chemical affinity', *Ambix*, **11**, 1963, pp. 55–64.

W. A. Smeaton, 'The Lunar Society and chemistry', *University of Birmingham Historical Journal*, **11**, 1967, pp. 144–54.

W. A. Smeaton, 'Louis Bernard Guyton de Morveau, F.R.S. (1737–1816) and his relations with British scientists', *Notes and Records of the Royal Society of London*, **22**, 1967, pp. 113–30.

W. A. Smeaton, 'E. F. Geoffroy was not a Newtonian chemist,' *Ambix*, **18**, 1971, pp. 212–4.

W. A. Smeaton and Bertel Linder, 'Schwediauer, Bentham, and Beddoes: translators of Bergman and Scheele', *Annals of Science*, **24**, 1968, pp. 259–73.

Barbara Smith and J. L. Moilliet, 'James Keir of the Lunar Society', *Notes and Records of the Royal Society*, **22**, 1967, pp. 144–54.

J. G. Smith, *The heavy chemical industries in France*, Oxford, 1979.

L. Stewart, *The rise of public science: rhetoric, technology and natural philosophy in Newtonian Britain, 1660–1750*, Cambridge, 1992.

J. M. Stillman, *The story of alchemy and early chemistry*, New York, 1960 (1924).

I. Strube, *Georg Ernst Stahl*, Leipzig, 1984.

R. Taton, *L'enseignement et diffusion des sciences en France au XVIIIe siècle*, Paris, 1964.

M. Teich and R. Young (eds) *Changing perspectives in the history of science*, London, 1973.

Arnold Thackray, '"Matter in a nut-shell": Newton's *Opticks* and eighteenth-century chemistry', *Ambix*, **15**, 1968, pp. 29–53.

Arnold Thackray, 'Quantified chemistry—the Newtonian dream', in D. S. L. Cardwell (ed.), *John Dalton and the progress of science*, Manchester, 1968, pp. 92–108.

Arnold Thackray, *Atoms and powers: an essay on Newtonian matter-theory and the development of chemistry*, Cambridge, Mass., 1970.

E. A. Underwood, 'English-speaking medical students at Leyden', *Nature*, **221**, 1969, pp. 810–14.

Paul Walden, 'The beginnings of the doctrine of chemical affinity', *Journal of Chemical Education*, **31**, 1954, pp. 27–33.

R. S. Westfall, *Science and religion in seventeenth-century England*, New Haven, 1958.

R. S. Westfall, 'The foundations of Newton's philosophy of nature', *British Journal for The History of Science*, **1**, 1962–3, pp. 171–82.

R. S. Westfall, *Force in Newton's physics*, London and New York, 1971.

R. S. Westfall, 'Newton and the Hermetic tradition', in A. G. Debus (ed.) *Science, medicine, and society in the Renaissance*, London, 1972, pp. 183–98.

R. S. Westfall, 'Isaac Newton's *Index chemicus*', *Ambix*, **22**, 1975, pp. 174–85.

R. S. Westfall, *Never at rest: a biography of Isaac Newton*, Cambridge, 1980.

R. S. Westfall, 'Alchemy in Newton's Library', *Ambix*, **31**, 1984, pp. 97–101.

L. L. Whyte (ed.), *Roger Joseph Boscovich*, London, 1961.

W. P. D. Wightman, 'William Cullen and the teaching of chemistry', *Annals of Science*, **11**, 1955, pp. 154–65 and **12**, 1956, pp. 192–205.

L. Pearce Williams, 'The politics of science in the French Revolution', with commentaries by H. B. Hill and H. Guerlac in the same volume, in Marshall Clagett (ed.), *Critical problems in the history of science*, Madison, 1959, pp. 291–308.

Rudolf Winderlich, 'Carl Friedrich Wenzel, 1740–93', *Journal of Chemical Education*, **27**, 1950, pp. 56–9.

A. Wolf, *A history of science, technology and philosophy in the eighteenth century*, 2nd edn, rev. by D. McKie, London, 1952.

Eri Yagi, 'Stephen Hales' work in chemistry: a Newtonian influence on eighteenth-century chemistry', *Japanese Studies in the History of Science*, **5**, 1966, pp. 75–86.

Index

ege Library